Letts

revise GCSE

Mark Patmore
and Brian Seager

Mathematics

Contents

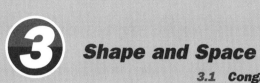

Shape and Space

Handling Data

This book and your GCSE course

STAY YOUR COURSE!   *Use these pages to get to know your course*

- Make sure you know your awarding body (exam board)
- Check which specification you are doing A, B or C

- Know how your course is assessed:
- what format are the papers?
- how is coursework assessed?
- how many papers?

	Syllabus number	Modular tests	Terminal papers	Coursework
AQA A	3301	None	Paper 1 no calculator 40% Paper 2 with calculator 40% F: $1\frac{1}{2}$ I: 2 H: 2 hours	20%
AQA B	3302	Module 1 : 25 minutes 11% Module 2 : 40 minutes 19% Section A with a calculator Section B without a calculator	Paper 1 no calculator 25% Paper 2 with calculator 25% F: 1 I: $1\frac{1}{4}$ H: $1\frac{1}{4}$ hours	20%
EDEXCEL A	1387	None	Paper 1 no calculator 40% Paper 2 with calculator 40% F: $1\frac{1}{2}$ I: 2 H: 2 hours	20%
EDEXCEL B	1388	Module 1 : 50 minutes 15% Module 2 : 50 minutes 15% Section A without calculator Section B with calculator	Paper in two sections Section A without calculator Section B with calculator All tiers 2 hours 50%	20%
OCR A	1962	None	Paper 1 no calculator 40% Paper 2 with calculator 40% F: $1\frac{1}{2}$ I: 2 H: 2 hours	20%
OCR B	1968	Note: Paper 1 may be taken early in the previous June or January	Papers in two sections Section A without calculator Section B with calculator Paper 1: $1\frac{1}{2}$ hours 30% Paper 2: 2 hours 50%	20%
OCR C	1966	At least two, 1 hour each 15% each Section A without calculator Section B with calculator	Paper in two sections Section A without calculator 25% Section B with calculator 25% All tiers 2 hours	20%
NICCEA		None	Paper 1 no calculator 40% Paper 2 with calculator 40% F: $1\frac{1}{2}$ I: 2 H: $2\frac{1}{2}$ hours	20%
WJEC	2210	None	Paper 1 no calculator 40% Paper 2 with calculator 40% F: $1\frac{1}{2}$ I: 2 H: $2\frac{1}{4}$ hours	20%

For all specifications the available grades are the same.
All specifications have the same content. For the Higher Tier, you need to cover all of this book. Contact your awarding body for full details of your course or download your complete GCSE specifications. www.aqa.org.uk, www.ocr.org.uk, www.edexcel.org.uk, www.wjec.co.uk, www.ccea.org.uk

Tier	Grades available
Foundation	G–D
Intermediate	E–B
Higher	C–A*

Preparing for the examination

Planning your study

The final three months before taking your GCSE examination are very important in achieving your best grade. However, your success can be assisted by adopting an organised approach throughout the course.

- After completing a topic in school or college go through the topic again in *Revise GCSE Maths Study Guide*. Copy out the main points, results and formulae, etc on a sheet of paper or use a highlighter pen to emphasise them.
- A few days later try to write out these key points again from memory. Check any differences between what you originally wrote and what you wrote later.
- If you have written your notes on a piece of paper keep this for revision later.
- Try some questions in the book and check your answers.
- Decide whether you have fully mastered the topic and write down any weaknesses you think you have.

Preparing a revision programme

At least three months before the final examination look at the list of topics in your awarding body's specification. Go through and identify which topics you feel you need to concentrate on. It is a temptation at this time to spend valuable revision time on the things you already know and can do. It makes you feel good but does not move you forward.

When you feel you have mastered all the topics spend time trying past questions. Each time check your answers with the answers given. In the final couple of weeks go back to your summary sheets (or your highlighting in the book).

How this book will help you

Revise GCSE Mathematics Guide will help you because:

- It contains the essential content for your GCSE course without the extra material that will not be examined.
- It contains Progress Checks and GCSE questions to help you confirm your understanding.
- It gives sample GCSE questions with model answers and advice from examiners on how to improve.
- The examination questions for 2003 are different from those in 2002 or 2001. Trying past questions will not help you when answering some parts of the questions in 2003. The questions in this book have been written by experienced examiners who are writing the questions for 2003 and beyond.
- The summary table and specification labels will give you a quick reference to the requirements for your examination.
- Marginal comments and highlighted key points will draw your attention to important things you might otherwise miss.

Five ways to improve your grade

1 Read the instructions carefully

Some of the instructions could be:

- Answer *ALL* the questions.
  This means answer as many as you can. Only by getting all of them right will you obtain full marks.

- Write your answers in the spaces provided on the question paper.
  You are not allowed to use any other paper.

- In the exam you should check that you have been given the correct paper, that you know how many questions you have to answer on that paper and how long you have to do it. Try to spread your time equally between the questions. If you do this it will avoid the desire to rush the paper or spend too much time on some questions and not finish the paper.

- Many mathematics papers start with fairly straightforward questions which may be shorter than those that follow. If this is the case work through them in order to build up your confidence. Do not overlook any part of a question and double check that you have seen everything on each paper, look especially at the back page in case there is a question there!

- Take time to read through all the questions carefully and then start with the question(s) that you think you can do best.

- When there are about 15 minutes remaining in the examination then quickly check if you are running out of time. If you think that you will run out of time then try to score as many marks as possible by concentrating on the easier parts, the first parts, of any questions that you have not yet attempted.

2 Read the question carefully

- Make sure you understand what the question is asking. Some questions are structured and some are unstructured – called 'multi-step' questions – and for these you will have to decide how to tackle the question and it would be worthwhile spending a few seconds thinking the question through.

- Make sure you understand key words. The following glossary may help you in answering questions:
  Write down, state – no explanation is needed for an answer
  Calculate, find, show, solve – include enough working to make your method clear
  Draw – plot accurately using the graph paper provided and selecting a suitable scale if one is not given. Such an instruction is usually followed by asking you to read one or more values from your graph.

- The number of marks is given in brackets [ ] at the end of each question or part question. This gives some indication of how many steps will be required to answer the question and therefore of what proportion of your time, you should spend on each part of the question.

3 Show your working and check your answers

- State units if required and give your final answer to an appropriate degree of accuracy.
- Write down the figures on your calculator and then make a suitable rounding. Don't round the numbers during the calculation. This will often result in an incorrect answer.
- Don't forget to check your answers, especially to see that they are reasonable. The mean height of a group of men will not be 187 metres!
- Lay out your working carefully and concisely. Write down the calculations you are going to make. You usually get marks for showing a correct method. (If you are untidy and disorganised, you might misread some of your own work and/or lose marks because the examiner cannot read your work or follow your method.)
- Remember that marks are given for the following:
    - using an appropriate method to answer a question
    - for facts found as you work through a question
    - for the final answer.
- Remember that if all that is written down is an answer and that answer is wrong you gain no marks. Once you have finished the paper if you have any time left check the work you have done. The best way to do this is to work through the questions again.

4 What examiners look for

The examiners look for the following:

- Work which is legible, clearly set out and easy to follow and understand. Use a pen, not pencil, except in drawings, and use the appropriate equipment.
- That drawings and graphs are neat, and graphs are labelled.
- That you always indicate how you obtain your answers.
- The right answer!

5 Practice makes perfect

- Practice all aspects of manipulative algebra, solving equations, rearranging formulas, expanding brackets, factorising, etc.
- Practise answering questions that ask for an explanation. Your answers should be concise and use mathematical terms where appropriate.
- Practise answering questions with more than one step to the answer, e.g. finding the radius of a sphere with the same volume as a given cone.
- Make sure you can use your calculator efficiently.

Coursework

For all GCSE Mathematics Specifications you will have to complete two coursework tasks, one on **Using and Applying Mathematics**, and one on **Handling Data**. Each task will contribute 10% towards your final grade leaving 80% of your marks to be earned from your performance on the written examination papers. Each task will be assessed using the appropriate criteria with marks being awarded for performance in 'assessment strands' which are outlined in the table below.

Most of the awarding bodies offer **two different methods** for assessing the two coursework tasks that are completed. These are:

- Your school can select the tasks to be completed, mark all of them and send a sample of marked scripts to the awarding body to be moderated.

Task type	Details	Weighting	Assessment strands	Max marks
Using and Applying Mathematics	One task, usually, based on number and/or algebra or based on shape and space.	10%	1 Making and monitoring decisions 2 Communicating mathematically 3 Developing the skills of mathematical reasoning	8 8 8
Handling data	One task (which should be based on a statistics activity rather than on a probability based activity)	10%	1 Specify and Plan 2 Collect, Process and Represent 3 Interpret and Discuss	8 8 8

- The awarding body will set the tasks and mark the work.

Thus the **maximum mark** for each task is **24**. The marks from each task that you gain will be added together to give a total coursework mark.

For example your marks could be:

	strand 1	strand 2	strand 3	Total
Using and Applying mathematics	6	6	5	17
Handling data	6	5	4	15
Final total				32

The detailed criteria for Using and Applying Mathematics and for Handling Data, which are common to all awarding bodies, are written by QCA and printed in the specification that you will be using. You should be able to obtain a copy of these detailed criteria from your teacher or the awarding body's website.

Coursework strategies

There are some well defined strategies that you might wish to adopt.

Using and Applying Mathematics Strand 1: Making and monitoring decisions

4 marks	Gathering (in a systematic manner) enough results that are correct and enable you to write a generalisation about the given problem.
5 marks	Change one variable and undertake sufficient new work so you could make a further generalisation.
6 marks	Show a range of techniques to extend and develop the task further. For example if you had only been using simple linear equations such as $y = 3x - 2$ up to this point you could try to use a graphical approach or simultaneous or quadratic equations to support this extended work. (This would link in with the requirement for 6 marks in the Communication strand where the consistent use of symbolism, i.e. algebra, is required).
7 marks	Attempt to co-ordinate three features in the work, perhaps by moving into 3 dimensions.

Using and Applying Mathematics Strand 2: Communicating mathematically

4 marks	Present work in an orderly manner using two different methods, for example tables and diagrams, linking them together with a commentary.
5–6 marks	Show an increased use of algebra.
7–8 marks	Show a sophisticated use of algebraic techniques.

Using and Applying Mathematics Strand 3: Mathematical reasoning

3 marks	Show a progression from 'making general statements', i.e. a valid generalisation, derived from at least three of your results.
4 marks	Test your findings, formula or relationship by checking a further case (do not use the values you already used in deriving the formula or results).
5 marks	Give a valid explanation as to why your generalisation works, referring to the shape of a grid, or size and structure of a shape.
6–8 marks	The progression continues up to 8 marks where a mathematically rigorous justification is expected.

Handling Data Strand 1: Specify and plan

5–6 marks	Show clear aims and state a plan designed to meet these aims. The data used should be appropriate and the reason for any sampling should be explained.
7–8 marks	Demonstrate valid reasons for what you have done and explain any limitations, for example bias, that might arise.

Handling Data Strand 2: Collect, process and represent	
5–6 marks	Show correct use of appropriate calculations using relevant data.
7–8 marks	Demonstrate evidence of higher level techniques applied accurately.

Handling Data Strand 3: Interpret and discuss	
5–6 marks	Use summary statistics to make comparisons between sets of data and clearly relating your findings back to the original problem and evaluating the success, (or otherwise), of your strategy.
7–8 marks	Explain how you avoided bias and demonstrate the use, for example, of a pre-test or a pilot questionnaire.

Tips

There are various techniques for you to use which should help you to improve your marks on both types of coursework:

One technique is to use a three part approach, and for each part think through some or all of the following questions, answering them in your head or on rough paper (but don't include them in your submission):

Part	Using and applying mathematics task	Handling data task
1 Starting a task	What does the task tell me?	Do I have a clear plan and clear aims?
	What does it ask me?	What questions do I have to answer? Is a hypothesis given or can I think of one?
	What can I do to get started? For example find a simple starting point	What data do I think I will need? When, where and how will I get it? – doing a survey, using a questionnaire, doing an experiment?
2 Working on the task	What connections are possible?	Am I using the right technique?
	Is my method clear?	Is my method clear?
	Is there a result to help me?	Is the data relevant and appropriate?
	Is there a pattern in the results I have found?	Are some patterns, trends and/or conclusions beginning to emerge?
	Can the problem be changed?	Can I think of different questions to ask?
3 The review, conclusion and extension	Is the solution acceptable?	Have I used the most appropriate presentation?
	Are my results presented as clearly as possible with explanations and reasons?	Are my results presented as clearly as possible with explanations and reasons
	Can it be extended?	Can I think of further questions to ask? – perhaps as a result of my findings so far?
	What conclusions can be made?	Have I answered the starting questions or hypothesis?

Overview

Topic	Section	Studied in class	Revised	Practice questions
1.1 Integers	Negative integers			
	Common factors and multiples			
	Prime numbers			
1.2 Powers and roots	Square roots and cube roots			
	Index laws			
	Inverse operations			
	Standard index form			
1.3 Fractions	Equivalent fractions			
	Adding and subtracting fractions			
	Multiplying fractions			
	Dividing fractions			
1.4 Decimals	Decimals and fractions			
1.5 Percentages	Percentages and fractions			
	Percentages and decimals			
	One quantity as a percentages of another			
	Finding a percentage of a quantity			
	Finding a percentage increase or decrease			
	Reversed percentages			
1.6 Ratio	Simplifying ratios			
	Dividing in a ratio			
1.7 Mental methods	Recall			
	Rounding numbers			
	Estimates			
1.8 Written methods	Proportional change			
	Irrational numbers			
1.9 Calculator methods	Understanding keys and display			
	Upper and lower bounds			
	Exponential growth and decay			
1.10 Solving problems	Strategies			
	Checking			

Number

1.1 Integers

LEARNING SUMMARY

After studying this section, you will be able to:

- **use and understand negative integers**
- **find common factors and multiples**
- **recognise and use prime numbers**

Negative integers

AQA A AQA B
EDEXCEL A EDEXCEL B
OCR A OCR B
OCR C
NICCEA
WJEC

Integers are whole numbers, positive and negative.

They can be represented on a number line:

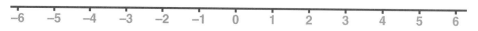

Moving (translating) to the right is adding, to the left subtracting.

A negative number is to the left of zero, a positive one to the right.

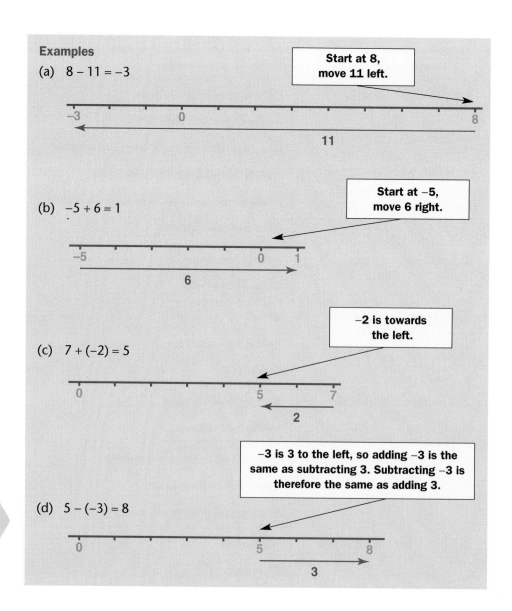

Examples

(a) $8 - 11 = -3$

> Start at 8, move 11 left.

(b) $-5 + 6 = 1$

> Start at −5, move 6 right.

(c) $7 + (-2) = 5$

> −2 is towards the left.

(d) $5 - (-3) = 8$

> −3 is 3 to the left, so adding −3 is the same as subtracting 3. Subtracting −3 is therefore the same as adding 3.

> You will be expected to do calculations like these without a calculator.

Common factors and multiples

This section concerns positive integers.

> **KEY POINT**
> If you can divide one number by another, the second is a **factor of the first**.

For example, **12** ÷ **3** = **4**, so **3** is a factor of **12** (and so are **4**, **2**, **6**, **12** itself and, of course **1**).

> **KEY POINT**
> If you multiply one number by another, the result is a **multiple of the first number** (and also of the second number).

For example, **24** is a multiple of **6** (since **6** × **4** = **24**). **24** is also a multiple of **2**, **3**, **4**, **8**, **12** and **24**.

Some numbers have **common** factors.

For example, **24** and **30** both have factors **2**, **3** and **6**.
The largest of these (**6**) is the **highest common factor (HCF)**.

Two (or more) numbers will have **common** multiples.

For example, **4** has multiples **4**, **8**, **12**, **16**, **20**, **24**, **28**, ...
3 has multiples **3**, **6**, **9**, **12**, **15**, **18**, **21**, **24**, **27**, ...
4 and **3** have common multiples **12**, **24**, ...
The smallest of these (**12**) is called the **lowest common multiple (LCM)**.

Prime numbers

All positive integers can be written as a **product of prime factors**.
For example, $11 = (1 \times)11$
$$12 = 2 \times 2 \times 3$$
$$39 = 3 \times 13$$

> **KEY POINT**
> Some numbers have no factor other than 1 and themselves. These are prime numbers. For example, 2, 11, 29 are prime numbers.

Examples

(a) Write 24 and 75 as a product of prime factors.
Hence find the LCM and HCF of 24 and 75.

2	24
2	12
2	6
3	3
	1

$24 = 2 \times 2 \times 2 \times 3$

3	75
5	25
5	5
	1

$75 = 3 \times 5 \times 5$

LCM must include $2 \times 2 \times 2 \times 3$ and $3 \times 5 \times 5$ but the 3 is repeated.

So LCM $= 2 \times 2 \times 2 \times 3 \times 5 \times 5 = 600$

> **You do not need a calculator to do these.**

The only common factor is 3, so 3 is the HCF.

(b) Find the LCM and HCF of 56 and 84.

2	56
2	28
2	14
7	7
	1

$56 = 2 \times 2 \times 2 \times 7$

2	84
2	42
3	21
7	7
	1

$84 = 2 \times 2 \times 3 \times 7$

> **What do you notice about the repeated factors and the HCF?**

LCM must include all factors but $2 \times 2 \times 7$ are repeated.

So LCM $= 2 \times 2 \times 2 \times 3 \times 7 = 168$

HCF $= 2 \times 2 \times 7 = 28$

 With practice, you can do these in your head but it is sensible to write down each step, as here. In an exam, you will be expected to do this sort of calculation without a calculator.

 PROGRESS CHECK

1 Work out:
(a) $5 + (-6) - (-3)$ (b) $-8 + 7 - (-9)$ (c) $2 + (-11) + (-3)$
2 Find (i) the prime factors (ii) the LCM and (iii) the HCF of these numbers.
(a) 27 and 36   (b) 75 and 60   (c) 18, 24 and 27.

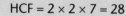

2 (a)(i) $27 = 3 \times 3 \times 3$; $36 = 2 \times 2 \times 3 \times 3$ (ii) 108 (iii) 9
(b)(i) $75 = 3 \times 5 \times 5$; $60 = 2 \times 2 \times 3 \times 5$ (ii) 300 (iii) 15
(c)(i) $18 = 2 \times 3 \times 3$; $24 = 2 \times 2 \times 2 \times 3$; $27 = 3 \times 3 \times 3$ (ii) 216 (iii) 3
1 (a) 2 (b) 8 (c) -12

1.2 Powers and roots

> **LEARNING SUMMARY**
>
> *After studying this section, you will be able to:*
> - **find square roots and cube roots**
> - **understand index laws**
> - **use inverse operations**
> - **use standard index form**

Square roots and cube roots

AQA A AQA B
EDEXCEL A EDEXCEL B
OCR A OCR B
OCR C
NICCEA
WJEC

> **KEY POINT**
> A square root of a number must be squared (that is multiplied by itself or raised to the power 2) to give the number.

Any positive number has two square roots, one positive and one negative.

For example, the square roots of **49** are **7** and **−7**.

> **KEY POINT**
> The cube root of a number must be cubed (that is raised to the power 3) to give the number.

> You can use your calculator to find square roots but it will only give the positive ones. Use the $\boxed{x^y}$ key to find the cube root.

For example, the cube root of **64** is **4**.

These can be written

$$\sqrt{49}, \quad \sqrt[3]{64}.$$

Index laws

AQA A AQA B
EDEXCEL A EDEXCEL B
OCR A OCR B
OCR C
NICCEA
WJEC

Using an index is a shorthand way to show multiplication and division.

For example, $2 \times 2 \times 2 = 2^3$, $\quad 3 \times 3 \times 3 \times 3 \times 3 = 3^5$

> Do not confuse 2^3 (= 8) with 2×3 (= 6)!

Examples

(a) Use indices to find 16×32.

$16 = 2 \times 2 \times 2 \times 2 \, (= 2^4)$, $32 = 2 \times 2 \times 2 \times 2 \times 2 \, (= 2^5)$,

$16 \times 32 = 2 \times 2 \times 2 \times 2 \times 2 \times 2 \times 2 \times 2 \times 2 \, (= 2^9)$

You can see that $2^4 \times 2^5 = 2^9$.

> **KEY POINT**
> This is an example of the index law for multiplication.
> $a^p \times a^q = a^{p+q}$

(b) Use indices to find $81 \div 9$.

$81 = 3 \times 3 \times 3 \times 3 \ (= 3^4), \ 9 = 3 \times 3 \ (= 3^2)$

$81 \div 9 = \dfrac{81}{9} = \dfrac{3 \times 3 \times 3 \times 3}{3 \times 3} = 3 \times 3$

This time $3^4 \div 3^2 = 3^{4-2} = 3^2$.

> **KEY POINT**
> This uses the index law for division:
> $a^p \div a^q = a^{p-q}$

(c) Find $(2^3)^4$.

This is

$(2^3) \times (2^3) \times (2^3) \times (2^3) = 2 \times 2 \times 2 \times 2 \times 2 \times 2 \times 2 \times 2 \times 2 \times 2 \times 2 \times 2$
$= 2^{12} = 2^{3 \times 4}$

> **Don't add the indices this time!**

> **KEY POINT**
> This shows another index law:
> $(a^p)^q = a^{p \times q}$

(d) Simplify $\dfrac{2^3 \times (3^2)^4}{2^2 \times 3^5}$.

This equals

> **You can cancel the factors that are common to the top and bottom of the fraction, using the index law for division.**

$$\dfrac{2^3 \times 3^8}{2^2 \times 3^5} = \dfrac{2^3}{2^2} \times \dfrac{3^8}{3^5} = 2 \times 3^3$$

(e) Use the index law for division to find $\dfrac{2^3}{2^3}$.

$$\dfrac{2^3}{2^3} = 2^{3-3} = 2^0$$

$$\text{But } \dfrac{2^3}{2^3} = 1, \text{ so } 2^0 = 1.$$

> **KEY POINT**
> This demonstrates another index law:
> $a^0 = 1$

(f) Use the index law for division to find $\dfrac{1}{3^2}$.

> $\dfrac{1}{a}$ is the reciprocal of a.

$$\dfrac{1}{3^2} = \dfrac{3^0}{3^2} = 3^{0-2} = 3^{-2}$$

> **KEY POINT**
> This is an example of the index law:
> $\dfrac{1}{a^p} = a^{-p}$

(g) If $\sqrt{2} = 2^x$, find the value of x.

$$(\sqrt{2})^2 = (2^x)^2$$

> **Square both sides:**
> $$(\sqrt{2})^2 = 2 = 2^1$$

$2 = 2^{2x}$, giving $2x = 1$ and $x = \frac{1}{2}$.

> **KEY POINT**
> **The general law is:**
> $$\sqrt[n]{a} = a^{\frac{1}{n}}$$

Inverse operations

AQA A AQA B
EDEXCEL A EDEXCEL B
OCR A OCR B
OCR C
NICCEA
WJEC

> **KEY POINT**
> **The operation which reverses what has been done is called an inverse operation.**

For example, The inverse of **adding 7** is **subtracting 7**;

the inverse of **dividing by 10** is **multiplying by 10**;

the inverse of **taking a reciprocal** is **taking the reciprocal again**;

> Remember that
> $\sqrt[n]{a} = a^{\frac{1}{n}}$.

the inverse of $\sqrt[n]{a}$ is $(\sqrt[n]{a})^n$, which equals a.

> $$\dfrac{1}{\frac{1}{a}} = a \text{ (Multiply top and bottom by } a.)$$

Standard index form

AQA A AQA B
EDEXCEL A EDEXCEL B
OCR A OCR B
OCR C
NICCEA
WJEC

There is a special form of index notation, which is very useful for showing large and small numbers.

For example,

$14\,000\,000 = 1.4 \times 10\,000\,000 = 1.4 \times 10 \times 10 \times 10 \times 10 \times 10 \times 10 \times 10$

$$= 1.4 \times 10^7$$

> **KEY POINT**
> **This is called standard index form. The number is expressed as a number between 1 and 10 multiplied by 10 to the appropriate index.**

> Find out how your calculator uses standard index form – how to enter numbers and how to read them. For example, your calculator may display 4.96^{05}, which you must write as 4.96×10^5. When keying numbers in, you do not usually have to enter the 10.
> (See Section 1.9, page 25)

Examples

(a) Write 2.7×10^4 as an ordinary number.

$$2.7 \times 10^4 = 2.7 \times 10 \times 10 \times 10 \times 10 = 27\,000$$

(b) Write 4.96×10^{-5} as an ordinary number.

$$4.96 \times 10^{-5} = 4.96 \div 10^5 = 0.000\,0496$$

Any of the three answers in this example is correct – unless the question asked for the answer in standard index form, when only 1.2×10^3 is right.

(c) Work out without using a calculator:

$$3 \times 10^6 \times 4 \times 10^{-4}$$

$$3 \times 10^6 \times 4 \times 10^{-4} = 3 \times 4 \times 10^{6-4} = 12 \times 10^2 = 1200 = 1.2 \times 10^3$$

(d) Work out, without using a calculator:

$$4.7 \times 10^4 + 8.2 \times 10^5$$

$$4.7 \times 10^4 + 8.2 \times 10^5 = 47\,000 + 820\,000 = 867\,000 = 8.67 \times 10^5$$

Unless you are asked to write down all the figures in your display (8.78561872 in this case), give your answer to sensible accuracy. See section 1.7, page 20.

(e) Use a calculator to find:

$$\frac{7.23 \times 10^5 \times 1.09 \times 10^{-3}}{8.97 \times 10^{-4}}$$

You should find the answer is 8.79×10^5, correct to two d.p.

Without using a calculator

1 Write as ordinary numbers:

(a) $\sqrt[3]{125}$ (b) 3^4 (c) $\sqrt{196}$ (d) 3.2×10^5 (e) $4^{-\frac{1}{2}}$

2 Simplify, leaving your answer in index form:

(a) $5^4 \times 5^3$ (b) $7^3 \div 7^5$ (c) $(3^2)^5$ (d) $\sqrt[3]{5}$ (e) $\dfrac{1}{13^4}$

3 Write in standard index form, as simply as possible:

(a) 0.021 (b) 48 700 (c) $\dfrac{1}{4}$ (d) 1.008 (e) $5 \times 10^3 \times 2 \times 10^6$

(f) $8 \times 10^2 \div (4 \times 10^3)$

PROGRESS CHECK

Using a calculator

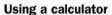

4 Work out the following, leaving your answers in standard index form:

(a) $3.7 \times 10^6 \times 1.02 \times 10^2$ (b) $4.5633 \times 10^8 \div (3.71 \times 10^2)$

(c) $1.57 \times 10^{-2} \times 3.81 \times 10^4$ (d) $7.81 \times 10^{-4} \div (9.734 \times 10^6)$

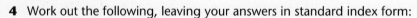

4 (a) 3.774×10^8 (b) 1.23×10^6 (c) 5.9817×10^2 (d) $8.02 \ldots \times 10^{-11}$
3 (a) 2.1×10^{-2} (b) 4.87×10^4 (c) 2.5×10^{-1} (d) 1.008 (e) 1.0×10^{10} (f) 2.0×10^{-1}
2 (a) 5^7 (b) 7^{-2} (c) 3^{10} (d) $5^{\frac{1}{3}}$ (e) 13^{-4}
1 (a) 5 (b) 81 (c) 14 (d) 320 000 (e) $\dfrac{1}{2}$ or 0.5

1.3 Fractions

LEARNING SUMMARY

After studying this section, you will be able to:

● find and use equivalent fractions
● add and subtract fractions
● multiply fractions
● divide fractions

Equivalent fractions

AQA A   AQA B
EDEXCEL A   EDEXCEL B
OCR A   OCR B
OCR C
NICCEA
WJEC

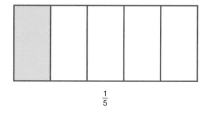

$\frac{1}{5}$

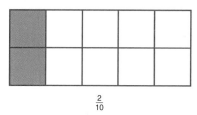

$\frac{2}{10}$

KEY POINT

If you multiply (or divide) the top (numerator) and the bottom (denominator) of a fraction by the same number, the value of the two fractions is the same. These are called **equivalent fractions**.

For example, $\dfrac{1}{5} = \dfrac{1 \times 2}{5 \times 2} = \dfrac{2}{10}$ and $\dfrac{18}{72} = \dfrac{18 \div 9}{72 \div 9} = \dfrac{2}{8} = \dfrac{2 \div 2}{8 \div 2} = \dfrac{1}{4}$

Using this, fractions can be put in order, simplified and added and subtracted.

Examples

(a) Write these fractions in order, smallest first.

$$\frac{2}{3}, \frac{3}{5}, \frac{4}{7}, \frac{13}{20}$$

> Find the LCM.
> 5 is a factor of 20.

The common denominator must have factors 3, 5, 7, 20.

LCM = $3 \times 5 \times 7 \times 4 = 420$

$$\frac{2}{3} = \frac{2 \times 5 \times 7 \times 4}{3 \times 5 \times 7 \times 4} = \frac{280}{420}, \qquad \frac{3}{5} = \frac{3 \times 3 \times 7 \times 4}{5 \times 3 \times 7 \times 4} = \frac{252}{420}$$

$$\frac{4}{7} = \frac{4 \times 3 \times 5 \times 4}{7 \times 3 \times 5 \times 4} = \frac{240}{420} \qquad \frac{13}{20} = \frac{13 \times 3 \times 7}{20 \times 3 \times 7} = \frac{273}{420}$$

> Make sure you multiply the numerator and denominator by the same numbers in each case.

If a calculator is allowed, then this is much more quickly done by turning the fractions into decimals. On a non-calculator paper you must use this method.

So the order is $\dfrac{4}{7}, \dfrac{3}{5}, \dfrac{13}{20}, \dfrac{2}{3}$

(b) Write as simply as possible: $\frac{160}{280}$

> This process is called cancelling.

$$\frac{160}{280} = \frac{160 \div 10}{28 \div 10} = \frac{16}{28} = \frac{16 \div 4}{28 \div 4} = \frac{4}{7}$$

Number

Adding and subtracting fractions

AQA A   AQA B
EDEXCEL A   EDEXCEL B
OCR A   OCR B
OCR C
NICCEA
WJEC

KEY POINT Only fractions with the same denominators can be added and subtracted.

Find the LCM of the denominators and change all fractions to equivalent fractions with that denominator.

> Expect these questions on the non-calculator paper.

Examples

(a) Find $\dfrac{3}{5} + \dfrac{2}{7}$.

LCM $= 5 \times 7 = 35$

$$\frac{3}{5} + \frac{2}{7} = \frac{3 \times 7}{5 \times 7} + \frac{2 \times 5}{7 \times 5} = \frac{21}{35} + \frac{10}{35} = \frac{31}{35}$$

> To find the number to multiply the numerator, divide the LCM by the denominator. Watch out for the subtraction.

(b) Find $\dfrac{2}{3} - \dfrac{5}{9} + \dfrac{3}{5}$.

LCM $= 9 \times 5 = 45$

$$\frac{2}{3} - \frac{5}{9} + \frac{3}{5} = \frac{2 \times 3 \times 5}{45} - \frac{5 \times 5}{45} + \frac{3 \times 9}{45} = \frac{30 - 25 + 27}{45} = \frac{32}{45}$$

Multiplying fractions

AQA A   AQA B
EDEXCEL A   EDEXCEL B
OCR A   OCR B
OCR C
NICCEA
WJEC

KEY POINT To multiply two fractions, multiply the numerators and multiply the denominators.

Examples

(a) Find $\dfrac{3}{4} \times \dfrac{5}{7}$

$$\frac{3}{4} \times \frac{5}{7} = \frac{3 \times 5}{4 \times 7} = \frac{15}{28}$$

> You can cancel the 5 and the 4 before multiplying.

(b) Work out $\dfrac{5}{8} \times \dfrac{4}{5}$

$$\frac{5}{8} \times \frac{4}{5} = \frac{5 \times 4}{8 \times 5} = \frac{20}{40} = \frac{1}{2}$$

(c) Work out $\dfrac{3}{5} \times 7$

> Any integer can be written with 1 as its denominator.

$$\frac{3}{5} \times 7 = \frac{3}{5} \times \frac{7}{1} = \frac{3 \times 7}{5 \times 1} = \frac{21}{5} = 4\frac{1}{5}$$

Dividing fractions

AQA A AQA B
EDEXCEL A EDEXCEL B
OCR A OCR B
OCR C
NICCEA
WJEC

A fraction represents a division.

For example, $\dfrac{1}{2} = 1 \div 2 = 0.5$

Multiplication and division are inverse operations (See page 17).

For example, dividing by **2** is the same as multiplying by $\dfrac{1}{2}$ and multiplying by **2** is the same as dividing by $\dfrac{1}{2}$.

Combining these ideas gives a method for dividing by a fraction.

> **Divide by 3 is the same as multiply by $\dfrac{1}{3}$.**
>
> **Divide by $\dfrac{1}{4}$ is the same as multiply by 4.**

For example, $\dfrac{5}{6} \div \dfrac{3}{4} = \dfrac{5}{6} \times \dfrac{4}{3} = \dfrac{5}{3} \times \dfrac{2}{3} = \dfrac{10}{9} = 1\dfrac{1}{9}$

> **KEY POINT**
>
> **To divide by a fraction, multiply by its reciprocal.**

Examples

> **This is the reciprocal of the example above.**

(a) Find $\dfrac{3}{4} \div \dfrac{5}{6}$

$$\dfrac{3}{4} \div \dfrac{5}{6} = \dfrac{3}{4} \times \dfrac{6}{5} = \dfrac{3}{2} \times \dfrac{3}{5} = \dfrac{9}{10}$$

> **Top-heavy fractions like $\dfrac{3}{2}$ are sometimes called improper fractions.**

(b) Work out $\dfrac{5}{8} \div 1\dfrac{1}{2}$

$$\dfrac{5}{8} \div 1\dfrac{1}{2} = \dfrac{5}{8} \div \dfrac{3}{2} = \dfrac{5}{8} \times \dfrac{2}{3} = \dfrac{5}{4} \times \dfrac{1}{3} = \dfrac{5}{12}$$

> **It sometimes comes as a surprise that dividing by a number less than 1 give a larger answer.**

(c) Find $15 \div \dfrac{3}{4}$

$$15 \div \dfrac{3}{4} = 15 \times \dfrac{4}{3} = 5 \times 4 = 20$$

> **PROGRESS CHECK**

Without using a calculator

1 Find the simplest fractions equivalent to these:

(a) $\dfrac{42}{63}$ (b) $\dfrac{34}{102}$ (c) $\dfrac{96}{144}$ (d) $\dfrac{78}{65}$

2 Work out:

(a) $\dfrac{3}{4} + \dfrac{3}{5}$ (b) $\dfrac{3}{5} + \dfrac{5}{6}$ (c) $\dfrac{7}{8} - \dfrac{2}{5}$ (d) $\dfrac{2}{3} - \dfrac{4}{7}$

3 Work out:

(a) $\dfrac{2}{3} \times \dfrac{5}{9}$ (b) $\dfrac{3}{8} \times \dfrac{4}{9}$ (c) $\dfrac{3}{4} \div \dfrac{1}{3}$ (d) $\dfrac{4}{5} \div \dfrac{2}{3}$

3 (a) $\dfrac{10}{27}$ (b) $\dfrac{1}{6}$ (c) $2\dfrac{1}{4}$ (d) $1\dfrac{1}{5}$

2 (a) $1\dfrac{7}{20}$ (b) $1\dfrac{13}{30}$ (c) $\dfrac{19}{40}$ (d) $\dfrac{2}{21}$

1 (a) $\dfrac{2}{3}$ (b) $\dfrac{1}{3}$ (c) $\dfrac{2}{3}$ (d) $\dfrac{6}{5}$ or $1\dfrac{1}{5}$

1.4 Decimals

After studying this section, you will be able to:

● *change a fraction into a decimal*
● *change a decimal into a fraction*

Decimals and fractions

AQA A   AQA B
EDEXCEL A   EDEXCEL B
OCR A   OCR B
OCR C
NICCEA
WJEC

KEY POINT

To change a fraction into a decimal, divide the numerator by the denominator.

Examples

(a) $\dfrac{2}{5} = 2 \div 5 = 0.4$

(b) $\dfrac{4}{7} = 4 \div 7 = 0.571\ 428\ 571\ \dots$

> The second decimal never stops. Can you see the repeating pattern?

KEY POINT

Decimals that never stop and have a repeating pattern are called recurring decimals.
All fractions give terminating or recurring decimals.

Example

Find decimals equivalent to these fractions:

$$\dfrac{13}{20}, \dfrac{2}{9}, \dfrac{5}{16}, \dfrac{5}{11}$$

$\dfrac{13}{20} = 13 \div 20 = 0.65$

$\dfrac{5}{16} = 0.3125$

$\dfrac{2}{9} = 0.222\ 222\ 222\ \dots$

$\dfrac{5}{11} = 0.454\ 545\ 454\ \dots$

> Which fractions give terminating decimals?

There is a simpler notation to show recurring decimals, using a dot above the number or numbers that make the pattern.

For example, $\dfrac{2}{9} = 0.222\ 222\ 222\ \dots = 0.\dot{2}$

$\dfrac{5}{11} = 0.454\ 545\ 45\ \dots = 0.\dot{4}\dot{5}$

> In this case, put a dot over the first and last figures in the pattern.

$\dfrac{4}{7} = 0.571\ 428\ 571\ \dots = 0.\dot{5}71\ 42\dot{8}$

Example

Find fractions equivalent to these decimals:

$0.7, 0.125, 0.04, 0.034, 0.\dot{3}, 0.\dot{2}\dot{7}, 0.0\dot{6}$

$0.7 = \dfrac{7}{10}$

$0.125 = \dfrac{125}{1000} = \dfrac{5 \times 5 \times 5}{10 \times 10 \times 10} = \dfrac{1}{8}$

> Cancel by 5 and by 5 and by 5.

$0.04 = \dfrac{4}{100} = \dfrac{1}{25}$

$0.034 = \dfrac{34}{1000} = \dfrac{17}{500}$

$0.\dot{3}$ needs a different approach.

$10 \times 0.\dot{3} = 3.333\,333\,33\,...$

$0.\dot{3} = 0.333\,333\,33\,...$

$10 \times 0.\dot{3} - 0.\dot{3} = 3$

> The two recurring parts are the same, so subtracting them gives 0.

$9 \times 0.\dot{3} = 3$, so $0.\dot{3} = \dfrac{3}{9} = \dfrac{1}{3}$

> This is the same method as before with 100 instead of 10 as there are two numbers in the pattern.

$100 \times 0.\dot{2}\dot{7} = 27.272\,727\,27\,...$

$0.\dot{2}\dot{7} = 0.272\,727\,27\,...$

$100 \times 0.\dot{2}\dot{7} - 0.\dot{2}\dot{7} = 27$

$99 \times 0.\dot{2}\dot{7} = 27$

$0.\dot{2}\dot{7} = \dfrac{27}{99} = \dfrac{3}{11}$

$10 \times 0.0\dot{6} = 0.666\,6\,...$

$0.0\dot{6} = 0.066\,66\,...$

$9 \times 0.0\dot{6} = 0.6$

$0.0\dot{6} = \dfrac{0.6}{9} = \dfrac{6}{90} = \dfrac{1}{15}$

Alternative method

$0.0\dot{6} = 0.\dot{6} \div 10 = \dfrac{2}{3} \div 10 = \dfrac{2}{30} = \dfrac{1}{15}$

1 Write as decimals:

(a) $\dfrac{4}{5}$ (b) $\dfrac{13}{40}$ (c) $\dfrac{2}{3}$ (d) $\dfrac{5}{9}$ (e) $\dfrac{17}{33}$

2 Write as fractions in their lowest terms:

(a) 0.15 (b) 0.046 (c) $0.\dot{7}$ (d) $0.0\dot{3}$ (e) $0.\dot{5}\dot{4}$

PROGRESS CHECK

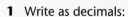

2 (a) $\dfrac{3}{20}$ (b) $\dfrac{23}{500}$ (c) $\dfrac{7}{9}$ (d) $\dfrac{1}{30}$ (e) $\dfrac{6}{11}$

1 (a) 0.8 (b) 0.325 (c) $0.\dot{6}$ (d) $0.\dot{5}$ (e) $0.\dot{5}\dot{1}$

 **Number**

1.5 Percentages

 LEARNING SUMMARY

After studying this section, you will be able to:

- *change percentages to fractions and vice versa*
- *change percentages to decimals and vice versa*
- *express one quantity as a percentage of another*
- *find a percentage of a quantity*
- *find a percentage increase or decrease*
- *use reversed percentages*

Percentages and fractions

AQA A   AQA B
EDEXCEL A   EDEXCEL B
OCR A   OCR B
OCR C
NICCEA
WJEC

A percentage is a fraction with denominator **100**. **51%** means $\dfrac{51}{100}$.

 KEY POINT   To change a percentage to a fraction, write it over 100 and cancel any common factors.

Example
Write as fractions:

20%, 45%, 66%, 140%.

$20\% = \dfrac{20}{100} = \dfrac{1}{5}$ →   **Cancelling by 20**

$45\% = \dfrac{45}{100} = \dfrac{9}{20}$ →   **Cancelling by 5**

Beware! This is NOT, $\frac{2}{3}$, which is $66\frac{2}{3}\%$

$66\% = \dfrac{66}{100} = \dfrac{33}{50}$ →   **Cancelling by 2**

$140\% = \dfrac{140}{100} = \dfrac{7}{5} = 1\frac{2}{5}$ →   **The result is larger than 1 as the percentage is larger than 100.**

KEY POINT   To change a fraction to a percentage, multiply by 100%.

Example
Change to percentages:

$\frac{2}{5}, \frac{5}{8}, 1\frac{1}{2}, \frac{1}{3}$

$\dfrac{2}{5} \times 100\% = \dfrac{2 \times 100}{5}\% = 2 \times 20\% = 40\%$

A percentage is the numerator of a fraction with 100 as denominator, so it is 100 times bigger than the fraction.

$\dfrac{5}{8} \times 100\% = \dfrac{5 \times 100}{8}\% = \dfrac{5 \times 25}{2}\% = 62\frac{1}{2}\%$

$$1\frac{1}{2} \times 100\% = 150\%$$

$$\frac{1}{3} \times 100\% = \frac{100}{3} = 33\frac{1}{3}\%$$

Or you could write 33.3̇%

Don't make the common mistake of confusing 30% with $\frac{1}{3}$.

Percentages and decimals

AQA A AQA B
EDEXCEL A EDEXCEL B
OCR A OCR B
OCR C
NICCEA
WJEC

KEY POINT To change a percentage to a decimal, write it over 100 and divide it out.

Example

Write as decimals:

17.5%, 84%, 250%, 0.2%

$$\frac{17.5}{100} = 0.175, \quad \frac{84}{100} = 0.84, \quad \frac{250}{100} = 2.5, \quad \frac{0.2}{100} = 0.002$$

KEY POINT To change a decimal to a percentage, multiply by 100%.

Examples

Write as percentages:

0.75, 0.29, 5.3, 0.0005.

0.75 = 0.75 × 100% = 75%, 0.29 = 0.29 × 100% = 29%,

5.3 = 5.3 × 100% = 530%, 0.0005 = 0.0005 × 100% = 0.05%

One quantity as a percentage of another

AQA A AQA B
EDEXCEL A EDEXCEL B
OCR A OCR B
OCR C
NICCEA
WJEC

KEY POINT To find one quantity as a percentage of another, divide the first by the second and multiply by 100%.

Examples

(a) What is 45 as a percentage of 150?

Cancel before multiplying if not using a calculator.

$$\text{Percentage} = \frac{45}{150} \times 100\% = \frac{3}{10} \times 100\% = 30\%$$

(b) Find 83 as a percentage of 745.

Round the answer to 3 significant figures.

$$\text{Percentage} = \frac{83}{745} \times 100\% = 11.14 ...\% = 11.1\%$$

 **Number**

Finding a percentage of a quantity

AQA A AQA B
EDEXCEL A EDEXCEL B
OCR A OCR B
OCR C
NICCEA
WJEC

KEY POINT To find a percentage of a quantity, multiply by the percentage and divide by 100.

> Cancel before multiplying if not using a calculator.

Examples

(a) Find 35% of 250.

$$\frac{35}{100} \times 250 = \frac{7}{20} \times 250 = \frac{7}{2} \times 25 = 87.5$$

(b) What is 6.5% of £23 500?

$$\frac{6.5}{100} \times 23\ 500 = £1527.50$$

Finding a percentage increase or decrease

AQA A AQA B
EDEXCEL A EDEXCEL B
OCR A OCR B
OCR C
NICCEA
WJEC

If a quantity is increased by a percentage, then that percentage of the quantity is added to the original.

For example,

if £50 is increased by **15%**, the result is **£50 + £$\frac{15}{100} \times 50$ = £57.50**

> This is much quicker than adding the increase.

This is the same as multiplying **50** by $1 + \frac{15}{100}$ or **1.15**.

KEY POINT To find the result of a percentage increase, multiply by (1 + the percentage divided by 100).

> To find a percentage increase, multiply by a number greater than 1.

Examples

(a) Increase 42 by 23%.

$\boxed{1 + \dfrac{23}{100} = 1.23}$

To increase by 23%, multiply by = 1.23

42 × 1.23 = 51.66

(b) The price of cars increased by 2.5%.

What is the price of a car previously costing £10 500?

To increase by 2.5%, multiply by = 1.025

$\boxed{1 + \dfrac{2.5}{100} = 1.025}$

10 500 × 1.025 = £10 762.50

Finding a decrease works in the same way. This time the percentage is subtracted.

For example, if **£50** is decreased by **15%**, the result is $£50 - £\dfrac{15}{100} \times 50$

$= £42.50$

This is the same as multiplying 50 by $1 - \dfrac{15}{100}$ or 0.85.

> **KEY POINT**
> To find the result of a percentage decrease, multiply by (1 – the percentage divided by 100).

To find a percentage decrease, multiply by a number smaller than 1.

Examples

(a) Decrease 68 by 35%.

To decrease by 35%, multiply by = 0.65

$1 - \dfrac{35}{100} = 0.65$

$68 \times 0.65 = 44.2$

(b) I bought a car for £8500.

A year later its value has fallen by 27%.

What is its value now?

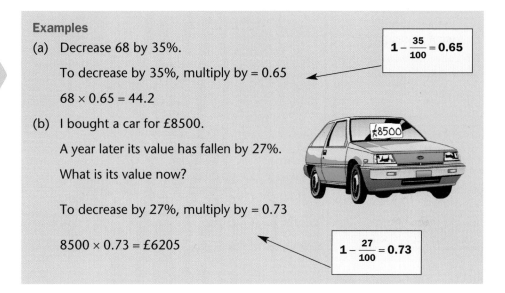

To decrease by 27%, multiply by = 0.73

$8500 \times 0.73 = £6205$

$1 - \dfrac{27}{100} = 0.73$

Reversed percentages

AQA A   AQA B
EDEXCEL A   EDEXCEL B
OCR A   OCR B
OCR C
NICCEA
WJEC

A washing machine costs £399.50, including VAT at 17.5%

 £399.50 inc VAT

If you think that decreasing by 17.5% gives the answer, try working it back again.
399.50 × 0.825 = £329.5875
329.5875 × 1.175 = £387.27!

What was the price without VAT?

The cost without VAT was multiplied by 1.175 to give £399.50.

Use the inverse operation. Divide £399.5 by 1.175, which gives £340.

> **KEY POINT**
>
> To find the value before a percentage increase, divide by (1 + the percentage divided by 100).
> To find the value before a percentage decrease, divide by (1 − the percentage divided by 100).

> The amount was multiplied by 1.05, so divide 472.50 by 1.05.

Examples

(a) What amount, when increased by 5% gives £472.50?

$$472.5 \div 1.05 = £450$$

> The amount was multiplied by 0.65, so divide 64.35 by 0.65.

(b) In a sale, all prices are reduced by 35%.

The sale price of a dress is £64.35. What was the original price?

$$64.35 \div 0.65 = £99$$

> **PROGRESS CHECK**

1 Change to fractions:
   (a) 30% (b) $33\frac{1}{3}$% (c) 85% (d) 210%

2 Change to decimals:
   (a) 20% (b) 43% (c) $17\frac{1}{2}$% (d) 155%

3 Change to percentages:
   (a) $\frac{3}{4}$ (b) $\frac{19}{20}$ (c) 0.37 (d) 0.015

4 Find:
   (a) 20 as a percentage of 50 (b) 7 as a percentage of 35
   (c) 18 as a percentage of 73

5 Find:
   (a) 18% of 73 (b) 6.8% of 142 (c) 0.45% of 100

6 Increase
   (a) 83 by 20% (b) 65 by 95% (c) 1890 by 7.5%

7 Decrease:
   (a) 47 by 70% (b) 195 by 17% (c) 4.7 by 5%

8 Find:
   (a) what was increased by $17\frac{1}{2}$% to give £28.20
   (b) what was increased by 6.5% to give £1597.50
   (c) what was decreased by 25% to give £12 375
   (d) what was decreased by 2.5% to give 83.85?

> You should attempt some of these questions without a calculator as practice for the examination

8 (a) £24 (b) £1500 (c) £16 500 (d) 86
7 (a) 14.1 (b) 161.85 (c) 4.465
6 (a) 99.6 (b) 126.75 (c) 2031.75
5 (a) 13.14 (b) 9.656 (c) 0.45
4 (a) 40% (b) 20% (c) 24.7%
3 (a) 75% (b) 95% (c) 37% (d) 1.5%
2 (a) 0.2 (b) 0.43 (c) 0.175 (d) 1.55
1 (a) $\frac{3}{10}$ (b) $\frac{1}{3}$ (c) $\frac{17}{20}$ (d) $2\frac{1}{10}$ or $\frac{21}{10}$

1.6 Ratio

> **LEARNING SUMMARY**
>
> After studying this section, you will be able to:
> - *simplify ratios*
> - *divide in a ratio*

Simplifying ratios

AQA A   AQA B
EDEXCEL A   EDEXCEL B
OCR A   OCR B
OCR C
NICCEA
WJEC

> **KEY POINT**   A ratio compares the relative sizes of two or more quantities.

This is usually written **5 : 2**

For example, the ratio of the number of days in the week to the number of days at the weekend is **5** to **2**.

Ratios behave in a similar way to fractions.

For example, I mix a drink with 40 ml concentrate and 240 ml water.

The ratio is **40 : 240 = 1 : 6**   ← **Divide each number by 40.**

Example
Write these ratios in their simplest form.

Divide each number in the ratio by the HCF.

2 : 8, 5 : 30, 20 : 70, 81 : 18

2 : 8 = 1 : 4, 5 : 30 = 1 : 6, 20 : 70 = 2 : 7, 81 : 18 = 9 : 2.

Dividing in a ratio

AQA A   AQA B
EDEXCEL A   EDEXCEL B
OCR A   OCR B
OCR C
NICCEA
WJEC

> **KEY POINT**   To divide in a ratio, add the parts of the ratio to give the denominators of the required fractions. Each part of the ratio gives the numerators and the resulting fractions are multiplied by the quantity to be divided.

For example, if **50** is to be divided in the ratio **3 : 7**, there are **3 + 7 = 10** parts.

The first share is therefore $\frac{3}{10}$ of the **50**, that is $\frac{3}{10} \times$ **50 = 15**.

Examples
(a) Divide 75 in the ratio 3 : 2.

$3 + 2 = 5$ so the fractions are $\frac{3}{5}$ and $\frac{2}{5}$.

$\frac{3}{5} \times 75 = 45$, $\frac{2}{5} \times 75 = 30$

Number

(b) Share £2500 in the ratio 1 : 3 : 4.

$1 + 3 + 4 = 8$ so the fractions are $\dfrac{1}{8}, \dfrac{3}{8}, \dfrac{4}{8}$.

The shares are

$\dfrac{1}{8} \times 2500 = £312.50, \dfrac{3}{8} \times 2500 = £937.50, \dfrac{1}{2} \times 2500 = £1250$

Always check that the parts correctly add to the total; £312.50 + £937.50 + £1250 = £2500.

(c) A sum of money is shared between Jane and Pat in the ratio 6 : 5 . Pat receives £250. How much does Jane receive?

£250 is 5 parts, so 6 parts will be $250 \times \dfrac{6}{5} = £300$

1 Simplify these ratios:
(a) 8 : 10   (b) 100 : 75   (c) $1\frac{1}{2} : 2\frac{1}{2}$ (d) 49 : 7 : 63

2 Share in the given ratio:
(a) 72 in the ratio 4 : 5   (b) 98 in the ratio 6 : 1
(c) £600 in the ratio 5 : 4 : 6

1 (a) 4 : 5 (b) 4 : 3 (c) 3 : 5 (d) 7 : 1 : 9
2 (a) 32, 40 (b) 84, 14 (c) £200, £160, £240

1.7 Mental methods

After studying this section, you will be able to:
● **recall number facts**
● **round numbers**
● **make estimates**

Recall

You should be able to recall many number facts, from easy ones like $8 + 7 = 15$ and $8 \times 7 = 56$, which you already know, to more difficult ones you may need to learn.

You should know:
● All integer squares from $2^2 = 4$ up to $15^2 = 225$
● The square roots of the perfect squares from 1 to 225
● Integer cubes $2^3 = 8$, $3^3 = 27$, $4^3 = 64$, $5^3 = 125$ and $10^3 = 1000$
● The cube roots of 8, 27, 64, 125 and 1000

AQA A, AQA B, EDEXCEL A, EDEXCEL B, OCR A, OCR B, OCR C, NICCEA, WJEC

Rounding numbers

AQA A AQA B
EDEXCEL A EDEXCEL B
OCR A OCR B
OCR C
NICCEA
WJEC

KEY POINT

Numbers are rounded for two main reasons:
- they are measurements and cannot be exact
- the accuracy of the discarded figures is not required.

For example, you measure a length of a straight line with a ruler and find that it is **5.8 cm**.

A more accurate measuring instrument might give a value **5.84 cm** but greater accuracy, such as **5.837 924**, could not be attempted.

The number **5.84** is correct to two decimal places.

Examples
Write these numbers to the given accuracy:

5 or larger in the next place, round up.

(a) 15.78 correct to one decimal place

(b) 0.0345 correct to three decimal places

(c) 45 291 correct to the nearest hundred

(a) 15.8   (b) 0.035   (c) 45300

Rounding to a number of decimal places is not the only way.
This is the same measurement, made to the same accuracy,

4236.7 mm, 423.67 cm, 4.2367 m, 0.004 236 7 km

They each have a different number of decimal places but the same number of **significant figures**, five.

KEY POINT

The number of significant figures is found by ignoring zeros, which merely denote the size of the number, and counting the other figures.

For example, **23.78** has four significant figures,

Zeros between non-zero digits are significant.

0.000 164 has three significant figures,
1.006 has four significant figures,
49 060 has four significant figures.

Example
Write these correct to the number of significant figures shown.

Don't be tempted to round part (c) in two stages – 1.0945 → 1.095 → 1.10.

In part (d), the zero is retained to show that it is correct to three significant figures.

(a) 65.794 (three)   (b) 0.003 86 (one)   (c) 1.0945 (three)
(d) 21.96 (three)

(a) 65.8   (b) 0.004   (c) 1.09   (d) 22.0

Estimates

AQA A   AQA B
EDEXCEL A   EDEXCEL B
OCR A   OCR B
OCR C
NICCEA
WJEC

KEY POINT

To estimate the result of a calculation, round or approximate each number so that you can work it out in your head. Rounding to one significant figure is often best.

Example

Estimate the answers to these calculations.

(a) $421 \times (73.6 - 21.7)$

(b) $\dfrac{0.0256 \times 937.2}{84.07 \times 0.567}$

(c) $(1.834 \times 10^{-4}) \div (9.7 \times 10^{5})$

'≈' means 'approximately equal'

(a) $421 \times (73.6 - 21.7) \approx 400 \times (70 - 20)$
$$= 400 \times 50$$
$$= 20\ 000$$

Further approximation during the calculation is sometimes needed

(b) $\dfrac{0.0256 \times 937.2}{84.07 \times 0.567} \approx \dfrac{0.03 \times 1000}{80 \times 0.6}$
$$= \dfrac{30}{48}$$
$$\approx \dfrac{3}{5}$$
$$= 0.6$$

(c) $(1.834 \times 10^{-4}) \div (9.7 \times 10^{5}) \approx (2 \div 10) \times (10^{-4} \div 10^{5})$
$$= 0.2 \times 10^{-9}$$
$$= 2 \times 10^{-10}$$

PROGRESS CHECK

1 State correct to the number of decimal places shown:
(a) 509.27 (one)   (b) 0.2739 (two)   (c) 43.797 (two)

2 State correct to the number of significant figures shown:
(a) 847.239 (two)   (b) 0.0736 (two)   (c) 80.174 (three)

3 Estimate to one significant figure:
(a) 983×23.1 (b) $0.0792 \div 4.21$ (c) $\dfrac{18.9 \times 53.2}{0.207}$
(d) $(2.7 \times 10^{3}) \times (8.16 \times 10^{-2})$

3 (a) 20 000 (b) 0.02 (c) 5000 (d) 2×10^{2}
2 (a) 850 (b) 0.074 (c) 80.2
1 (a) 509.3 (b) 0.27 (c) 43.80

1.8 Written methods

After studying this section, you will be able to:

LEARNING SUMMARY

- calculate proportional change
- recognise irrational numbers

Proportional change

AQA A AQA B
EDEXCEL A EDEXCEL B
OCR A OCR B
OCR C
NICCEA
WJEC

KEY POINT Proportional change is when a quantity is increased or decreased in a given ratio. This is achieved by using a multiplier.

For example, percentage increase is a proportional change.

Increasing by **50%** means multiplying by $1 + \dfrac{50}{100} = 1.5$.

This is the same as increasing in the ratio **1.5 : 1** or **3 : 2**.

Turn the ratio into something : 1
The left-hand side of the ratio will give the multiplier.

It is better to use a fraction in this case to avoid awkward decimals.

Examples

(a) Increase 38 in the ratio 5 : 2.

Ratio 5 : 2 = 2.5 : 1, so multiplier is 2.5.

$38 \times 2.5 = 95$

(b) Decrease 900 in the ratio 2 : 9.

Ratio 2 : 9 = $\dfrac{2}{9}$: 1, so multiplier is $\dfrac{2}{9}$.

$900 \times \dfrac{2}{9} = 200$

Irrational numbers

AQA A AQA B
EDEXCEL A EDEXCEL B
OCR A OCR B
OCR C
NICCEA
WJEC

KEY POINT A rational number is one that can be written as a fraction with numerator and denominator both integers.
An irrational number is one that is not rational, so cannot be written as a fraction.

Examples of rational numbers are **5** $\left(= \dfrac{5}{1}\right)$, **0.75** $\left(= \dfrac{3}{4}\right)$, $\dfrac{5}{11}$.

Examples of irrational numbers are $\sqrt{2}$, $5 + 2\sqrt{3}$, π.

Numbers like $5 + 2\sqrt{3}$ are called surds.

Are numbers like $\dfrac{1}{\sqrt{3}}$ and $\dfrac{5}{5+2\sqrt{3}}$ irrational?

Examples

Multiplying top and bottom by $\sqrt{3}$ does not alter the size of the number.

(a) Show that $\dfrac{1}{\sqrt{3}}$ is an irrational number.

$$\frac{1}{\sqrt{3}} \times \frac{\sqrt{3}}{\sqrt{3}} = \frac{\sqrt{3}}{3} \text{ which is irrational.}$$

(b) Show that $\dfrac{5}{5+2\sqrt{3}}$ is an irrational number.

Expand these brackets:

$$(5 + 2\sqrt{3})(5 - 2\sqrt{3}) = 25 + 10\sqrt{3} - 10\sqrt{3} - 4\sqrt{3}\sqrt{3} = 25 - 12 = 13$$

For a better understanding of why this works, see Chapter 2, section 1

$$\frac{5(5 - 2\sqrt{3})}{(5 + 2\sqrt{3})(5 - 2\sqrt{3})} = \frac{25 - 10\sqrt{3}}{13}$$

Multiply top and bottom by $5 - 2\sqrt{3}$.

and this is an irrational number.

(c) $\dfrac{1 + \sqrt{2}}{1 - \sqrt{2}}$

$$\frac{(1 + \sqrt{2})(1 + \sqrt{2})}{(1 - \sqrt{2})(1 + \sqrt{2})} = \frac{3 + 2\sqrt{2}}{1 - 2}$$

Multiply top and bottom by $1 + \sqrt{2}$.

$$= -(3 + 2\sqrt{2})$$

and this is an irrational number.

---

1 Change in the given ratio:
(a) 18, 4 : 3
(b) 500, 3 : 4
(c) 0.6, 5 : 4

2 Show whether these numbers are rational or irrational:

(a) $\sqrt{3} \times \sqrt{6}$

(b) $\dfrac{1}{\sqrt{5}} \times \sqrt{20}$

(c) $\dfrac{1}{5 + 2\sqrt{2}}$

(d) $\dfrac{4}{\sqrt{2} - 1}$

2 (a) $3\sqrt{2}$, irrational (b) 2, rational (c) $\dfrac{5 - 2\sqrt{2}}{17}$ irrational (d) $4 + 4\sqrt{2}$, irrational

1 (a) 24 (b) 375 (c) 0.75

1.9 Calculator methods

LEARNING SUMMARY

After studying this section, you will be able to:

● *understand keys and display*
● *find upper and lower bounds*
● *understand exponential growth and decay*

Understanding keys and display

> **KEY POINT**
>
> Calculators are not all the same. You must learn how yours works.

You will also need sin, cos and tan for Chapter 3 and statistics keys for Chapter 4.

For this section you need to use:

● **memory keys**

● **brackets**

● **power key** (sometimes labelled 'x^y') for proportional change and exponential growth.

● **the exponent key** (sometimes labelled 'EXP') to give standard index form and also interpret the display.

Upper and lower bounds

> **KEY POINT**
>
> Measurements are not exact.
> When a number is stated to a certain accuracy,
> the greatest it could be is the **upper bound**
> and the least it could be is the **lower bound**.

For example, 234 mm is correct to the nearest millimetre.

The upper bound is **234.5** and the lower bound is **233.5**, since if it were bigger than **234.5** it would be recorded as **235**.

Example

(a) The sides of a rectangle are measured as 12.7cm and 35.8 cm, both to the nearest 0.1 cm. Find the upper and lower bounds of the area of the rectangle.

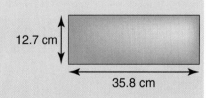

Bounds for the measurements are 12.65, 12.75, 35.75 and 35.85. The upper bound for the area will come from multiplying the upper bounds $12.75 \times 35.85 = 457.0875 \text{ cm}^2$

Similarly the lower bound is $12.65 \times 35.75 = 452.2375 \text{ cm}^2$

(b) A car travels 5000 m, correct to the nearest metre, in 2 minutes 21.7 seconds, correct to the nearest tenth of a second.

Find the upper and lower bounds of the average speed in metres per second.

Bounds for the measurements are 4999.5, 5000.5, 141.65, 141.75.

The upper bound for the speed will come from the upper bound for the distance divided by the lower bound for the time,
$5000.5 \div 141.65 = 35.3018$ m/s.

> **The most accurate statement of the speed is 35.3 m/s**

Similarly the lower bound is $4999.5 \div 141.75 = 35.2698 \dots$ m/s

Exponential growth and decay

AQA A AQA B
EDEXCEL A EDEXCEL B
OCR A OCR B
OCR C
NICCEA
WJEC

> **KEY POINT**
>
> Exponential describes a situation when the variable is in the index.

For example, If the number of bacteria doubles every hour,
– then after one hour, multiply the starting number by **2**
– after 2 hours, multiply the starting number by $2 \times 2 = 2^2$
and so on.

This also applies to repeating proportional changes or percentage increases/decreases.

For example, **£5000** is invested at **6.5%** compound interest for **5** years.

> With compound interest, the interest is added on and earns interest the next year(s).

After one year the amount is 5000×1.065.
After two years the amount is 5000×1.065^2
After five years the amount is $5000 \times 1.065^5 = £6850.43$

> **This is where you need the power key on your calculator!**

> **PROGRESS CHECK**
>
> 1 Find the upper and lower bounds for $\dfrac{85.7}{183 \times 0.27}$.
>   Each number is correct to the number of figures shown.
> 2 A town has 47 000 inhabitants. It is growing by a factor of 1.2 a year.
>   How many inhabitants will there be in 10 years time?
> 3 A substance of mass 200 g is decaying at a rate such that it will be $\dfrac{9}{10}$ of its mass at the end of the day.
>   What will be the mass after 15 days?
>
> **3** 41.2 g (to 3 significant figures)
> **2** 291 000 (to nearest 100)
> **1** 1.773 …, 1.697 …

1.10 Solving problems

Strategies

Many problems are set in a way that the mathematics you need to solve them is clear – so long as you know it!

Others, sometimes called multi-step or unstructured, are not so obvious and may involve several steps to be identified. All GCSE papers now contain some of these.

Before starting a problem like this, consider all the methods you know which may be relevant and then select the most appropriate.

Checking

It is important to check your work for mistakes.

It is especially true for answers derived from the calculator.

You do not need to do all of these every time.

KEY POINT

Always check a calculator answer by
- **considering if it is reasonable**
- **doing it again**
- **starting with your answer, use inverse operations to go back to the starting number.**
- **doing an approximate calculation.**

PROGRESS CHECK

1. (a) How many steps of 900 mm must you make in one minute to walk at 7 km/hour?
   (b) At what speed will you walk if the step is shortened by 25 mm but you take the same number of steps in a minute?
2. A shopkeeper makes a profit of 40% on his cost prices.
   He increases his selling prices by 10% but allows a 5% discount for cash.
   (His cost prices stay the same.)
   What is his percentage profit on cash sales?

2 46.3%
1 (a) 130 (b) 6.8(25) km/hour

Sample GCSE questions

1 Mr Smith starts a business with £50 000.
Three months later, Mr Jones joins the business. He invests £25 000.

The profit at the end of the year is £11 000. It is divided in proportion to how much was invested and for how long. **[5]**

How much does each receive?

> *Mr Smith invests £50 000 for 12 months.*
> *Mr Jones invests £25 000 for 9 months.*

Divide by 25 000 to simplify.

> *Ratio = 50000 × 12 : 25000 × 9* ✔✔
>   *= 24 : 9*
>   *= 8 : 3* ✔

8 + 3 = 11

> *£11 000 is divided in ratio 8 : 3.*
> *Mr S has $\frac{8}{11} \times 11\cdot000 = £8000$* ✔

Check the total is £11 000.

> *Mr J has $\frac{3}{11} \times 11\cdot000 = £3000$* ✔

2 PQRS is a rectangular piece of paper, size A0.

$PQ : PS = \sqrt{2} : 1$.
The area of PQRS is 1 m².

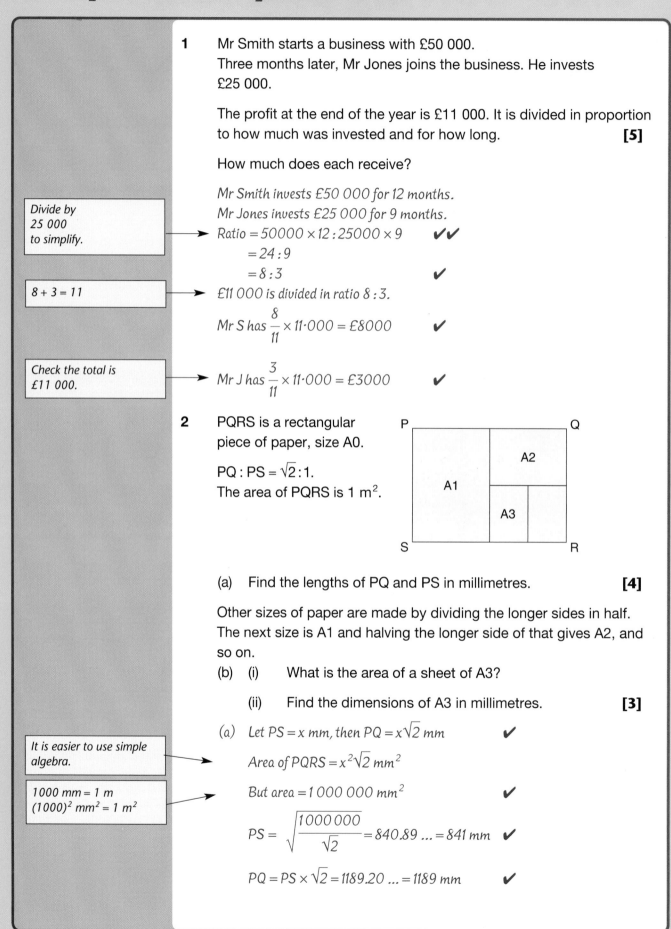

(a) Find the lengths of PQ and PS in millimetres. **[4]**

Other sizes of paper are made by dividing the longer sides in half. The next size is A1 and halving the longer side of that gives A2, and so on.

(b) (i) What is the area of a sheet of A3?

(ii) Find the dimensions of A3 in millimetres. **[3]**

> *(a) Let PS = x mm, then PQ = x√2 mm* ✔

It is easier to use simple algebra.

> *Area of PQRS = x²√2 mm²*

1000 mm = 1 m (1000)² mm² = 1 m²

> *But area = 1 000 000 mm²* ✔
>
> *PS = $\sqrt{\dfrac{1\,000\,000}{\sqrt{2}}} = 840.89 ... = 841$ mm* ✔
>
> *PQ = PS × √2 = 1189.20 ... = 1189 mm* ✔

Sample GCSE questions

Use all the digits, not previously rounded answers.

(b) (i) Area $= \frac{1}{8}$ PQRS $= 125\,000$ mm^2 ✔

(ii) Dimensions of A3 are $\frac{1}{2}$ PS, $\frac{1}{4}$ PQ

$= 420$ mm and 297 mm ✔✔

Check that these multiply together to ≈ 125 000.

3 The number of members in a club decreases by 5% every year. This year there are 120 members.

(a) How many will there be in 10 years time? **[3]**

(b) How many were there five years ago? **[2]**

It is appropriate to round down. Can you see why?

(a) 5% decrease gives multiplier $1 - 0.05 = 0.95$ ✔

$120 \times (0.95)^{10}$ ✔

$= 71.84 ... = 71$ ✔

Don't be tempted to multiply by $(1.05)^5$. Division is the inverse of multiplication.

(b) $120 \div (0.95)^5$ ✔

$= 155.08... = 155$ ✔

Check this answer by multiplying by $(0.95)^5$ to give 120.

4 Two accounts each contain £3000 earning compound interest. The interest rates are:

• Account A, 5% per annum
• Account B, 6% per annum (first two years), 4% per annum (third and subsequent years).

After four years, which account contains more and what is the difference? **[6]**

Multiplier 1.05 for four years.

Account A: $3000 \times (1.05)^4$ ✔

$= £3646.52$ ✔

First two years.

Account B: $3000 \times (1.06)^2$ ✔

Second two years.

$\times (1.04)^2$ ✔

$= £3645.86$ ✔

Account A contains more by 66p. ✔

5 Evaluate (do not use a calculator):

(a) $(27^{\frac{1}{3}})^2$ **[1]**

(b) $(1\frac{9}{16})^{-\frac{1}{2}}$ **[2]**

(c) $\frac{2}{3}\left(\frac{4}{7} - \frac{2}{5}\right)$ **[3]**

Sample GCSE questions

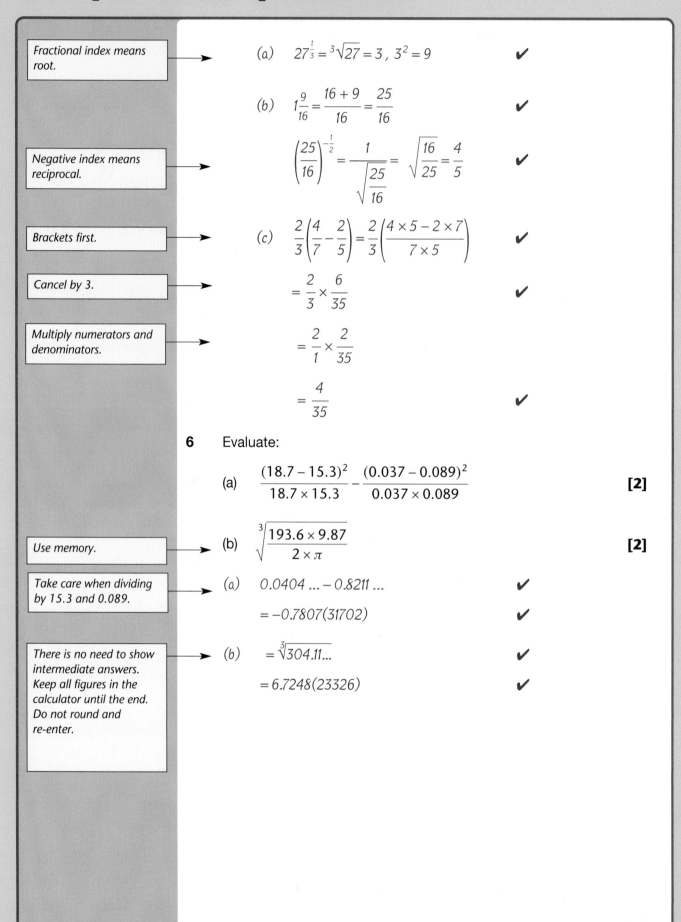

Fractional index means root.

(a) $27^{\frac{1}{3}} = \sqrt[3]{27} = 3$, $3^2 = 9$ ✔

(b) $1\frac{9}{16} = \frac{16 + 9}{16} = \frac{25}{16}$ ✔

Negative index means reciprocal.

$\left(\frac{25}{16}\right)^{-\frac{1}{2}} = \frac{1}{\sqrt{\dfrac{25}{16}}} = \sqrt{\dfrac{16}{25}} = \dfrac{4}{5}$ ✔

Brackets first.

(c) $\dfrac{2}{3}\left(\dfrac{4}{7} - \dfrac{2}{5}\right) = \dfrac{2}{3}\left(\dfrac{4 \times 5 - 2 \times 7}{7 \times 5}\right)$ ✔

Cancel by 3.

$= \dfrac{2}{3} \times \dfrac{6}{35}$ ✔

Multiply numerators and denominators.

$= \dfrac{2}{1} \times \dfrac{2}{35}$

$= \dfrac{4}{35}$ ✔

6 Evaluate:

(a) $\dfrac{(18.7 - 15.3)^2}{18.7 \times 15.3} - \dfrac{(0.037 - 0.089)^2}{0.037 \times 0.089}$ **[2]**

Use memory.

(b) $\sqrt[3]{\dfrac{193.6 \times 9.87}{2 \times \pi}}$ **[2]**

Take care when dividing by 15.3 and 0.089.

(a) $0.0404 \ldots - 0.8211 \ldots$ ✔

$= -0.7807(31702)$ ✔

There is no need to show intermediate answers. Keep all figures in the calculator until the end. Do not round and re-enter.

(b) $= \sqrt[3]{304.11\ldots}$ ✔

$= 6.7248(23326)$ ✔

Exam practice questions

1 (a) Write 1170 as a product of prime factors. **[3]**

(b) Simplify the following.

(i) $4^{-\frac{1}{2}} \times 8^{\frac{2}{3}}$ (ii) $\dfrac{\sqrt{3}}{9^{\frac{3}{4}}}$ **[4]**

(c) Find the value of this expression. Give your answer in standard index form.
$3 \times 10^{-2} + (4 \times 10^{-3}) \times (2 \times 10^{2})$ **[3]**

2 (a) Find the fractions equivalent to these decimals. Write each fraction in its lowest terms.
(i) 1.15 (ii) $1.\dot{1}\dot{5}$ **[5]**

(b) Simplify the following, showing whether each is rational or irrational.

(i) $(1 + \sqrt{2})^2$ (ii) $(1 + \sqrt{2})(1 - \sqrt{2})$ (iii) $\dfrac{1}{1 + \sqrt{2}}$ **[7]**

3 (a) A litre of petrol costs 79.9p, including VAT at 17.5%.
What is the cost before VAT is added?   **[3]**

(b) Last year, petrol cost 69.9p (including VAT).

(i) What was the percentage increase?   **[2]**
(ii) If this rate of growth continues, how much will a litre of petrol cost in five years' time?   **[3]**

4 These data are from a car journey.

Average speed   90 km per hour (to nearest 5 kph)
Time taken   2 hours (to nearest 0.1 hour)
Fuel used   23 litres (to nearest litre)

Find:
(a) the maximum value of the distance travelled   **[3]**

(b) the least value of the rate of fuel use (in litres per km).   **[4]**

5 Calculate the following, giving yours answers as fractions in their lowest terms.

(a) $1\frac{2}{5} + 3\frac{3}{4}$ **[3]**

(b) $1\frac{2}{5} \times 3\frac{3}{4}$ **[3]**

(c) $3\frac{1}{3} - 2\frac{5}{6}$ **[2]**

2 Algebra

Overview

Topic	Section	Studied in class	Revised	Practice questions
2.1 Symbols	**Letter symbols**			
	Manipulation			
	Know the words			
2.2 Index notation	**Index laws**			
2.3 Equations	**Setting up equations**			
2.4 Linear equations	**Solving linear equations**			
2.5 Formulae	**Substituting into formulae**			
	Changing subject of formulae			
	Generating formulae			
2.6 Direct and inverse proportion	**Solving problems**			
	Graphical interpretation			
2.7 Simultaneous linear equations	**Solving simultaneous linear equations**			
	Finding the solution on a graph			
2.8 Inequalities	**Inequalities with one variable**			
	Inequalities with two variables			
2.9 Quadratic equations	**Solving quadratic equations by factorising**			
	Solving quadratic equations by completing the square			
	Solving quadratic equations by using the formula			
2.10 Simultaneous linear and quadratic equations	**Solving simultaneous equations when one equation is quadratic**			
2.11 Numerical methods	**Trial and improvement**			
2.12 Sequences	**Generating sequences**			
	Finding the nth term			
2.13 Graphs of linear functions	**Equations of straight lines**			
	Parallel and perpendicular lines			
2.14 Interpreting graphical information	**Graphs of real-life situations**			
2.15 Quadratic and other functions	**Graphical solutions**			
	Graphs of functions			
2.16 Transformation of functions	**Transforming graphs**			
2.17 Loci	**Constructing loci**			
	Graphs of circles			

2.1 Symbols

LEARNING SUMMARY

After studying this section, you will be able to:

● *use letters as symbols*
● *manipulate algebraic expressions*
● *understand the meaning of the words used*

Letter symbols

AQA A   AQA B
EDEXCEL A   EDEXCEL B
OCR A   OCR B
OCR C
NICCEA
WJEC

KEY POINT

Letters are used as symbols to represent:

unknown number(s) in an equation, which can be found, for example x in $x^2 + 3x + 2 = 0$
variables in formulae, which can take many values, for example $v = u + at$
numbers in an identity, which can take any values, for example $5(x - 2) \equiv 5x - 10$, for any value of x

Manipulation

AQA A   AQA B
EDEXCEL A   EDEXCEL B
OCR A   OCR B
OCR C
NICCEA
WJEC

The rules for manipulating algebra are much like those in arithmetic.

For example,
$2a + 5a = 7a$, $b \times b = b^2$,
$c^3 \div c^2 = c$,
$2d(3d - 7) = 6d^2 - 14d$.

> **Remember the conventions:**
> $y \times 2 = 2y$ rather than
> $y2$, which could be
> confused with y^2,
> coming from $y \times y$.

This leads to more complicated examples.

> **Every term in one bracket must be multiplied by every term in the other. You can only collect like terms.**

Examples
$(a + b)(c + d) = a(c + d) + b(c + d) = ac + ad + bc + bd$
$(x - 5)(x + 2) = x(x + 2) - 5(x + 2) = x^2 + 2x - 5x - 10 = x^2 - 3x - 10$

This process can be reversed. It is called factorising.

For example,
$12y^2 - 6y = 6y(2y - 1)$ ◄———

> **Each term can be divided by $6y$, so $6y$ is a factor.**

Example
$x^2 + 5x + 6 = (x + 2)(x + 3)$

How did that happen?

You could work it out by trial and error, using your experience of expanding brackets.

A more systematic approach will save time!

Look at what happens when the brackets are expanded.

$(x + 2)(x + 3) = x^2 + 3x + 2x + 6$.

If it was given in this form you could spot the factors.

$x(x + 3) + 2(x + 3)$, which is $(x + 2)(x + 3)$.

$\boxed{2 + 3 = 5, \; 2 \times 3 = 6}$

Can you see how the 2 and the 3 combined to give 5 and 6?

> **KEY POINT**
> To factorise $x^2 + px + q$, find two numbers that will add to give p and multiply to give q.

The quadratic expressions will not always have plus signs.

For example,

$y^2 - 6y + 9$

The **+9** could result from multiplying two positive numbers but what about the **−6**?

Of course, **9** also equals **−3 × −3** or **−1 × −9**. Try these.

$\boxed{\begin{array}{l} -3 + -3 = -6, \; -3 \times -3 = 9 \\ y^2 - 6y + 9 = (y - 3)(y - 3) = (y - 3)^2 \end{array}}$

> **KEY POINT**
> To factorise $x^2 + px + q$, find two numbers that will add to give p and multiply to give q.
> This still works if p is a negative number.

Now try to factorise $x^2 - 3x - 10$.

Start with the −10. It could be -1×10, 1×-10, -2×5, 2×-5.

Notice that the signs are different, one positive and other negative.

Try the first:

$(x - 1)(x + 10) = x^2 - x + 10x - 10$

$\boxed{2 + -5 = -3, \; 2 \times -5 = -10}$

The two x terms have a different sign so they are subtracted. $10 - 1 = 9$, $1 - 10 = -9$, $-2 + 5 = 3$, so

$x^2 - 3x - 10 = (x + 2)(x - 5)$

> **KEY POINT**
> To factorise $x^2 + px + q$, when q is negative and p is positive or negative, find two numbers that will add to give p and multiply to give q.
> This still works so long as you remember the signs.

Know the words

AQA A   AQA B
EDEXCEL A   EDEXCEL B
OCR A   OCR B
OCR C
NICCEA
WJEC

These words are used in algebra:

Expression – any arrangement of letter symbols and possibly numbers.

Formula – connects two expressions containing variables, the value of one variable depending on the values of the others. It must have an equals sign.

Equation – connects two expressions involving definite unknown quantities. It also has an equals sign.

Identity – connects expressions involving unspecified numbers. An identity remains true whatever numerical values replace the letter symbols. It has a '≡' sign.

Function – a relationship between two sets of values such that a value from the first set maps on to a unique value in the second set.

PROGRESS CHECK

1 Simplify:

(a) $7a - 4b + 2a + 5b$ (b) $\dfrac{6ab}{2b^2}$ (c) $2x^2 + 3xy - 4y^2 + 2xy - 3x^2 + y^2$

2 Expand the brackets and simplify, if possible.

(a) $3x(5x - 7)$ (b) $(y + 4)(y - 3)$ (c) $(2x - 7)(5x - 3)$

3 Factorise:

(a) $x^2 + 7x + 12$ (b) $x^2 - 8x + 16$ (c) $y^2 - y - 20$ (d) $y^2 - 36$

3 (a) $(x + 3)(x + 4)$ (b) $(x - 4)^2$ (c) $(y - 5)(y + 4)$ (d) $(y - 6)(y + 6)$

2 (a) $15x^2 - 21x$ (b) $y^2 + y - 12$ (c) $10x^2 - 41x + 21$

1 (a) $9a + b$ (b) $\dfrac{3a}{b}$ (c) $5xy - x^2 - 3y^2$

2.2 *Index notation*

After studying this section, you will be able to:

● *do multiplication and division and find roots using indices*

Index laws

AQA A AQA B
EDEXCEL A EDEXCEL B
OCR A OCR B
OCR C
NICCEA
WJEC

The index laws are the same as in Chapter 1.

Here is a reminder.

KEY POINT

The index law for multiplication and division: $a^p \times a^q = a^{p+q}$, $a^p \div a^q = a^{p-q}$.
The index law for brackets: $(a^p)^q = a^{p \times q}$. The law for zero index: $a^0 = 1$.
The reciprocal index law: $\dfrac{1}{a^p} = a^{-p}$. The law for roots is: $\sqrt[n]{a} = a^{\frac{1}{n}}$.

PROGRESS CHECK

1 Simplify:

(a) $(a^2 b^4)^{\frac{1}{2}}$ (b) $(x^3 y^7)^0$ (c) $(a^{\frac{1}{3}} b^{\frac{3}{4}}) \times (a^{\frac{2}{3}} b^{\frac{1}{4}})$

2 Simplify:

(a) $(x^{-3} y^4) \div (x^{-2} y^2)$ (b) $\dfrac{p^3 q^{-4}}{p^{-1} q^{-2}}$ (c) $\sqrt{\dfrac{a^5 b^6}{a^7}}$

2 (a) $x^{-1} y^2$ or $\dfrac{y^2}{x}$ (b) $p^4 q^{-2}$ or $\dfrac{p^4}{q^2}$ (c) $\dfrac{b^3}{a}$ or $b^3 a^{-1}$

1 (a) ab^2 (b) 1 (c) ab or $a^{\frac{1}{1}}b$ or $a\sqrt{b}$

Algebra

2.3 Equations

LEARNING SUMMARY

After studying this section, you will be able to:

- *create an equation using symbols from given information*
- *set up an equation*

Setting up equations

AQA A AQA B
EDEXCEL A EDEXCEL B
OCR A OCR B
OCR C
NICCEA
WJEC

The information you are given will include an unknown quantity. Unless you are told otherwise, state which letter you will use to represent this quantity.

For example,

The angles round a point are $x°$, $2x°$, $(x + 50)°$ and $(x - 30)°$.

This time x is given.

The sum of the angles round a point is equal to 360°, so the equation is

$x + 2x + (x + 50) + (x - 30) = 360$, which simplifies to $5x = 340$

> Leave out the degree symbol as both sides are in degrees.

PROGRESS CHECK

1 Find and simplify these equations.
  (a) The angles of a triangle are $y°$, $3y°$, $(60 - y)°$
  (b) The perimeter of a rectangle is 32 cm.
      One side is three times as long as another.
  (c) Brian is three times as old as Adrian.
      In five years' time, Brian will be twice as old as Adrian.

1 (a) $y + 3y + (60 - y) = 180$, $3y = 120$
    (b) Let shorter side be x cm: $2x + 6x = 32$, $8x = 32$
    (c) Let Brian be b years old and Adrian a years: $b = 3a$, $b + 5 = 2(a + 5)$

2.4 Linear equations

LEARNING SUMMARY

After studying this section, you will be able to:

- *solve simple linear equations*
- *solve more complicated linear equations*

Solving linear equations

AQA A AQA B
EDEXCEL A EDEXCEL B
OCR A OCR B
OCR C
NICCEA
WJEC

KEY POINT

An equation has two parts separated by an equals sign. The arithmetic operations you perform on an equation must be the same for each part.

Some equations are very simple.

Always do the same operation on both sides of the equation.

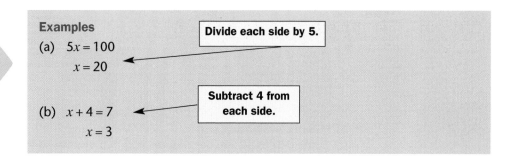

Examples

(a) $5x = 100$

$x = 20$

> Divide each side by 5.

(b) $x + 4 = 7$

$x = 3$

> Subtract 4 from each side.

Some equations involve both the previous ideas.

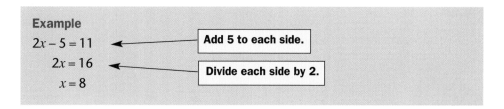

Example

$2x - 5 = 11$ ⟵ Add 5 to each side.

$2x = 16$ ⟵ Divide each side by 2.

$x = 8$

More complicated equations can involve brackets, with the unknown on one side or both.

Check first to see if each side has a common factor. If it does, divide each side by the factor.

Don't forget the negative signs.

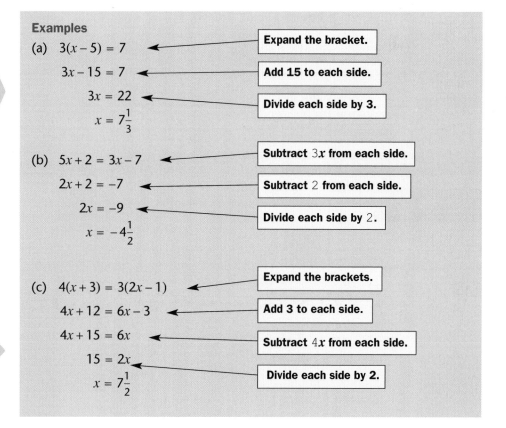

Examples

(a) $3(x - 5) = 7$ ⟵ Expand the bracket.

$3x - 15 = 7$ ⟵ Add 15 to each side.

$3x = 22$ ⟵ Divide each side by 3.

$x = 7\frac{1}{3}$

(b) $5x + 2 = 3x - 7$ ⟵ Subtract $3x$ from each side.

$2x + 2 = -7$ ⟵ Subtract 2 from each side.

$2x = -9$ ⟵ Divide each side by 2.

$x = -4\frac{1}{2}$

(c) $4(x + 3) = 3(2x - 1)$ ⟵ Expand the brackets.

$4x + 12 = 6x - 3$ ⟵ Add 3 to each side.

$4x + 15 = 6x$ ⟵ Subtract $4x$ from each side.

$15 = 2x$ ⟵ Divide each side by 2.

$x = 7\frac{1}{2}$

Some of the previous equations had solutions involving fractions. Some equations have fractions in them.

It is much easier to remove the fractions first, by multiplying each side by the lowest common multiple of the denominators of the fractions.

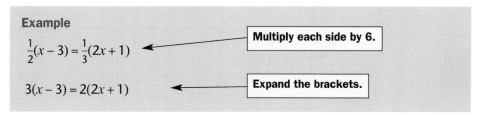

Example

$\frac{1}{2}(x - 3) = \frac{1}{3}(2x + 1)$ ⟵ Multiply each side by 6.

$3(x - 3) = 2(2x + 1)$ ⟵ Expand the brackets.

Algebra

$$3x - 9 = 4x + 2$$ ← Subtract $3x$ from each side.

$$-9 = x + 2$$ ← Subtract 2 from each side.

$$-11 = x \text{ or } x = -11$$

PROGRESS CHECK

1. Solve these equations.
   (a) Your equation in question 1(a) in the previous Progress Check (page 46).
   (b) Your equation in question 1(b) in the previous Progress Check.
2. (a) $3(x+4) = 16$ (b) $2 - 5x = x + 14$ (c) $2(5x+1) = 7x - 6$
3. (a) $\frac{1}{2}x = 40$ (b) $\frac{1}{4}(3x-5) = 2x+3$ (c) $3y + \frac{1}{2} = 4 + \frac{2}{3}y$

3. (a) $x = 80$ (b) $x = -3\frac{2}{5}$ (c) $y = 1\frac{1}{2}$
2. (a) $x = 1\frac{1}{3}$ (b) $x = -2$ (c) $x = -2\frac{2}{3}$
1. (a) $y = 40$ (b) $x = 4$

2.5 Formulae

LEARNING SUMMARY

After studying this section, you will be able to:

● *substitute numbers in formulae*
● *change the subject of a formula*
● *generate a formula*

Substituting into formulae

AQA A   AQA B
EDEXCEL A   EDEXCEL B
OCR A   OCR B
OCR C
NICCEA
WJEC

Replace the letters in the formula with the given numbers then do the arithmetic.

Example

$$s = ut + \frac{1}{2}at^2$$

Find the value of s when $u = 10$, $t = 5$ and $a = 0.27$.

$$s = 10 \times 5 + 0.5 \times 0.27 \times 5^2$$
$$= 50 + 3.375$$
$$= 53.375$$

Changing the subject of formulae

AQA A   AQA B
EDEXCEL A   EDEXCEL B
OCR A   OCR B
OCR C
NICCEA
WJEC

KEY POINT   When manipulating formulae, the rules are the same as for equations.

In the formula $s = ut + \frac{1}{2}at^2$, s is called the **subject** of the formula, since the formula is arranged to give s immediately on substitution. Changing the subject means rearranging the formula so that it has a different letter as the subject.

Examples
(a) Make u the subject of the formula

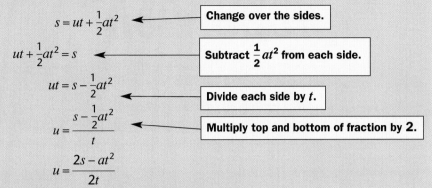

$$s = ut + \frac{1}{2}at^2$$ — Change over the sides.

$$ut + \frac{1}{2}at^2 = s$$ — Subtract $\frac{1}{2}at^2$ from each side.

$$ut = s - \frac{1}{2}at^2$$ — Divide each side by t.

$$u = \frac{s - \frac{1}{2}at^2}{t}$$ — Multiply top and bottom of fraction by **2**.

$$u = \frac{2s - at^2}{2t}$$

> When the new subject appears more than once, make sure you collect those terms. Don't leave the 'subject' somewhere else in the formula.

(b) Make a the subject of the formula

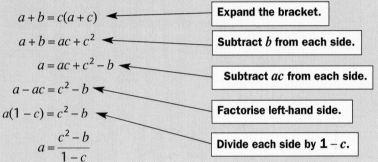

$$a + b = c(a + c)$$ — Expand the bracket.
$$a + b = ac + c^2$$ — Subtract b from each side.
$$a = ac + c^2 - b$$ — Subtract ac from each side.
$$a - ac = c^2 - b$$ — Factorise left-hand side.
$$a(1 - c) = c^2 - b$$ — Divide each side by $1 - c$.
$$a = \frac{c^2 - b}{1 - c}$$

Generating formulae

AQA A   AQA B
EDEXCEL A   EDEXCEL B
OCR A   OCR B
OCR C
NICCEA
WJEC

Instead of being given a formula, you may be asked to find one.

Example
A rectangle has perimeter P and one side is length L. Find a formula for its area A.
The area of a rectangle is its length multiplied by its width.
The length is L. The length plus the width is half the perimeter, so
width $= \frac{1}{2}P - L$.
So the formula is
$$A = L\left(\frac{1}{2}P - L\right)$$

PROGRESS CHECK

1 Substitute in these formulae.
   (a) Find v when $v = u + at$, and $u = 50$, $a = 10$ and $t = 2$.
   (b) Find s when $s = ut - \frac{1}{2}at^2$ and $u = 0$, $a = -6$ and $t = 10$.
   (c) Find v when $v^2 = u^2 - 2as$ and $u = 13$, $a = 6$ and $s = 12$.
2 (a) Make a the subject of the formula $v = u + at$.
   (b) Make P the subject of the formula $A = L(\frac{1}{2}P - L)$.
   (c) Make t the subject of the formula $3t - s = t(s - 6)$.
3 (a) Find a formula for the area A of a circle with circumference C.
   (b) Find a formula for the surface area A of a cube with volume V.

3 (a) $A = \dfrac{C^2}{4\pi}$ (b) $A = 6V^{\frac{2}{3}}$

2 (a) $a = \dfrac{v-u}{t}$ (b) $P = \dfrac{2(A+L^2)}{L}$ (c) $t = \dfrac{s}{9-s}$

1 (a) $v = 70$ (b) $s = 300$ (c) $v = 5$ (or -5)

Algebra

2.6 Direct and inverse proportion

> **LEARNING SUMMARY**
>
> *After studying this section, you will be able to:*
> - *solve problems involving proportion*
> - *represent proportional relationships on a graph*

Solving problems

 AQA A   AQA B
 EDEXCEL A EDEXCEL B
OCR A OCR B
OCR C
NICCEA
WJEC

> **KEY POINT**
>
> The symbol used to indicate proportion is ∝. So if y is directly proportional to x, write $y \propto x$.
> This may also be expressed as an equation $y = kx$, where k is a constant.

Other proportions

y is inversely proportional to x: $y \propto \dfrac{1}{x}$

> *This is the range of what you are expected to know.*

y is inversely proportional to the square of x: $y \propto \dfrac{1}{x^2}$

y is proportional to the square of x: $y \propto x^2$

y is proportional to the square root of x: $y \propto \sqrt{x}$

To solve a problem involving proportion, there are two main approaches. One is to change the relationship into an equation.

Example

The strength of a radio signal s is inversely proportional to the square of the distance d from the source. The strength is 8 when the distance is 5.

What is the strength when the distance is 50?

> *Using an equation*

$s \propto \dfrac{1}{d^2}$, giving $s = \dfrac{k}{d^2}$.

When $d = 5$, $s = 8$, so $8 = \dfrac{k}{25}$, giving $k = 200$.

When $d = 50$, $s = \dfrac{200}{50^2} = 0.08$

Another approach is to use multipliers.

For example, with the same problem.

Using multipliers $s \propto \dfrac{1}{d^2}$, and d increases from **5** to **50**, that is by a multiplier of $50 \div 5 = 10$.

The multiplier for s is $\dfrac{1}{(\textbf{Multiplier for } d)^2} = \dfrac{1}{100}$

New value for $s = 8 \times \dfrac{1}{100} = \textbf{0.08}$, as before.

Graphical interpretation

AQA A AQA B
EDEXCEL A EDEXCEL B
OCR A OCR B
OCR C
NICCEA
WJEC

KEY POINT

> The graph representing a particular proportional relationship will be in the shape of the graph for the corresponding equation. The graph for y directly proportional to x will be $y = kx$.

Other proportion graphs:

y is inversely proportional to x: $y = \dfrac{k}{x}$

y is inversely proportional to square of x: $y = \dfrac{k}{x^2}$

y is proportional to the square of x: $y = kx^2$

y is proportional to the square root of x: $y = k\sqrt{x}$

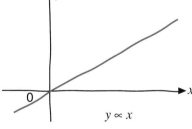
$y \propto x$

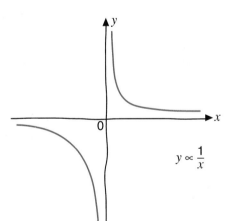
$y \propto \dfrac{1}{x}$

Sketch graphs like these do not need scales – or rulers!

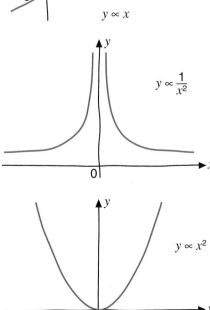
$y \propto \dfrac{1}{x^2}$

$y \propto x^2$

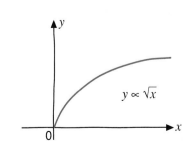
$y \propto \sqrt{x}$

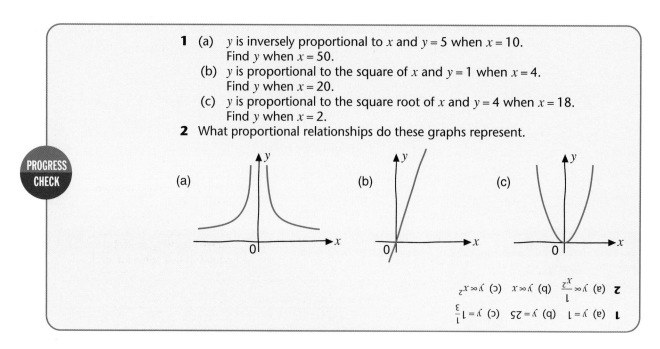

1 (a) y is inversely proportional to x and $y = 5$ when $x = 10$.
Find y when $x = 50$.
(b) y is proportional to the square of x and $y = 1$ when $x = 4$.
Find y when $x = 20$.
(c) y is proportional to the square root of x and $y = 4$ when $x = 18$.
Find y when $x = 2$.
2 What proportional relationships do these graphs represent.

(a)

(b)

(c)

2 (a) $y \propto \dfrac{1}{x^2}$ (b) $y \propto x$ (c) $y \propto x^2$

1 (a) $y = 1$ (b) $y = 25$ (c) $y = 1\dfrac{1}{3}$

2.7 Simultaneous linear equations

After studying this section, you will be able to:

- *solve simultaneous linear equations by elimination*
- *solve simultaneous linear equations by substitution*
- *interpret the solutions graphically in terms of straight lines*

'Simultaneous' means 'taken together'.

Solving simultaneous linear equations

AQA A   AQA B
EDEXCEL A   EDEXCEL B
OCR A   OCR B
OCR C
NICCEA
WJEC

Two equations in two unknowns can have a unique solution. The first method to find this solution is called **elimination**.

Example

Solve

$x + 2y = 7$

$4x - 3y = 6$

The object is to make either the x terms the same or the y terms the same. Multiply the first equation by 4.

$4x + 8y = 28$

$4x - 3y = 6$

Subtract the second equation from your new equation.

Remember, when you subtract −3 it is the same as adding 3.

$\overline{11y = 22}$

Divide each side by 11.

$y = 2$

$x + 2 \times 2 = 7$

Substitute for y in the first equation.

$x = 3$

The solution is $x = 3$, $y = 2$.

The simultaneous equations can also be solved by writing one of the equations in the form '$x = ...$' or '$y = ...$' This is the second method and it is called substitution.

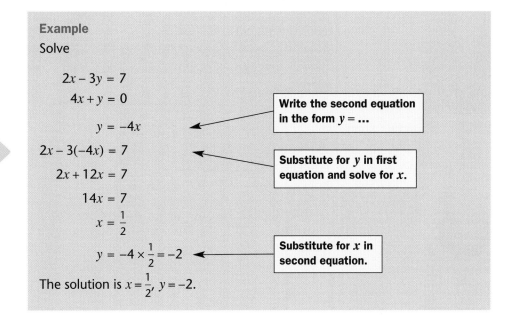

Example

Solve

$$2x - 3y = 7$$
$$4x + y = 0$$
$$y = -4x$$

Write the second equation in the form $y = ...$

$$2x - 3(-4x) = 7$$
$$2x + 12x = 7$$
$$14x = 7$$
$$x = \frac{1}{2}$$

Substitute for y in first equation and solve for x.

$$y = -4 \times \frac{1}{2} = -2$$

Substitute for x in second equation.

The solution is $x = \frac{1}{2}$, $y = -2$.

Take care with minus signs.

Finding the solution on a graph

AQA A AQA B
EDEXCEL A EDEXCEL B
OCR A OCR B
OCR C
NICCEA
WJEC

An equation like $2x + 3y = 5$ can be represented on a graph. It will be a straight line (which is why it is called a linear equation).

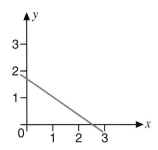

The coordinates of any point on the line will satisfy the equation, that is the equation will be true when the coordinates are substituted for x and y.

The point (1, 1) is on the line and $2 \times 1 + 3 \times 1 = 5$.

Look at two straight lines drawn on the same axes. The equation of the other line is $x - 2y = -1$.

There is more about drawing graphs in Section 2.13 on page 64.

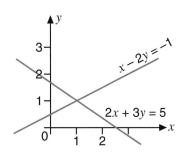

The point where they meet has coordinates that satisfy both equations.

> **KEY POINT**
>
> The coordinates of the point of intersection of two straight lines give the solution to the corresponding simultaneous equations.

This will work both ways round:

Draw the graphs to solve the simultaneous equations.

Find the point where two straight lines meet by solving their equations simultaneously.

PROGRESS CHECK

1 Solve these pairs of simultaneous equations:
   (a) $x + 2y = 1$, $2x + y = 5$ (b) $x + 2y = 6$, $3x - 6y = 12$
   (c) $3x - 2y = -9$, $4x + 5y = -\frac{1}{2}$
2 Draw graphs to solve these equations:
   (a) $x + y = 6$, $y = 2x$ (b) $x - y = -4$, $x + y = 2$
3 Use algebra to find the points of intersection of these lines:
   (a) $4x - y = 5$, $3x - 2y = 0$ (b) $6x - y = 4$, $4x + y = 1$

3 (a) $(2, 3)$ (b) $(\frac{1}{2}, -1)$
2 (a) $x = 2$, $y = 4$ (b) $x = -1$, $y = 3$
1 (a) $x = 3$, $y = -1$ (b) $x = 5$, $y = \frac{1}{2}$ (c) $x = -2$, $y = 1\frac{1}{2}$

2.8 Inequalities

LEARNING SUMMARY

After studying this section, you will be able to:

● *solve inequalities with one variable*
● *solve inequalities with two variables*

Inequalities with one variable

AQA A	AQA B
EDEXCEL A	EDEXCEL B
OCR A	OCR B
OCR C	
NICCEA	
WJEC	

Example

Solve the inequality

$3x - 5 > 2(x - 2)$.   ← | **Expand the bracket.** |

$3x - 5 > 2x - 4$ ← | **Subtract $2x$ from each side.** |

$x - 5 > -4$ ← | **Add 5 to each side.** |

$x > 1$

> **KEY POINT**
>
> The rules for manipulating inequalities are like those for equations.

The next example is different.

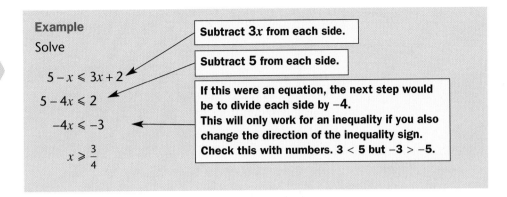

Example

Solve

$$5 - x \leqslant 3x + 2$$

$$5 - 4x \leqslant 2$$

$$-4x \leqslant -3$$

$$x \geqslant \frac{3}{4}$$

> Subtract $3x$ from each side.

> Subtract 5 from each side.

> If this were an equation, the next step would be to divide each side by -4.
> This will only work for an inequality if you also change the direction of the inequality sign.
> Check this with numbers. $3 < 5$ but $-3 > -5$.

> Sometimes the inequality includes equals. It is solved in the same way.

> **KEY POINT**
> Manipulate inequalities like equations except that when multiplying or dividing each side by a negative number, change the direction of the inequality sign.

Inequalities with two variables

AQA A   AQA B
EDEXCEL A   EDEXCEL B
OCR A   OCR B
OCR C
NICCEA
WJEC

Here is an inequality in two variables.

$$2x - y > 3$$

This is similar to the equations in the previous section. The graph of $2x - y = 3$ is a straight line.

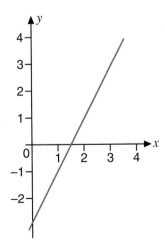

> If you write the equation in the form $y = 2x - 3$ you can see that it crosses the y-axis at -3 and has gradient 2. (See Section 2.13.)

What is the value of $2x - y$ at a point above the line, say $(1, 4)$?
It is -2. Try a point below the line, say $(4, -2)$. Here $2x - y = 10$.
Try other points and you will find that on the line $2x - y = 3$.
Above the line, $2x - y < 3$.
Below the line $2x - y > 3$.

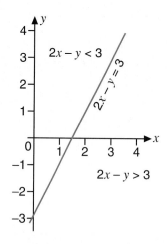

The solution to the inequality $2x - y > 3$ is a region. This is shown **unshaded** on the diagram.

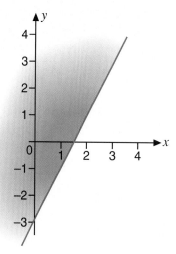

Several linear inequalities can be solved simultaneously.

Example

Show on a graph the solution set of the inequalities $2x - y > 3$, $x + y < 5$ and $y > 0$.

Shade out the unwanted regions.

A quick way to test which side of the line satisfies the inequality is to substitute $(0, 0)$. It won't work for $y > 0$ though!

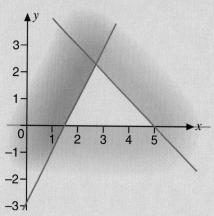

The solution set is the unshaded triangle.

1 Solve these inequalities:
   (a) $2(3x - 7) > 5x - 3$ (b) $4 - 3x \leqslant 2x + 7$ (c) $3(3 - x) < 5(5 - x)$
2 Show the solution set for these inequalities on a graph:
   $x + y \leqslant 2$, $y - 2x \leqslant 5$, $x \leqslant 0$, $y \geqslant 0$

1 (a) $x > 11$ (b) $x \geqslant -0.6$ (c) $x > 8$
2 Quadrilateral with vertices $(0, 0)$, $(0, 2)$, $(-1, 3)$, $(-2\frac{1}{2}, 0)$

2.9 Quadratic equations

LEARNING SUMMARY

After studying this section, you will be able to:

- *solve quadratic equations by factorising*
- *solve quadratic equations by completing the square*
- *solve quadratic equations by using the formula*

A quadratic equation is one that contains a term with the variable squared.

Solving quadratic equations by factorising

AQA A   AQA B
EDEXCEL A   EDEXCEL B
OCR A   OCR B
OCR C
NICCEA
WJEC

If the quadratic equation is on the non-calculator paper, it is worth looking for factors.

Quadratic expressions were factorised in Section 2.1 on page 43.
For an equation, the first step is to write it in the form $ax^2 + bx + c = 0$.
Then try to find the factors of the left-hand side.

Examples

(a) Solve $x^2 - 3x + 2 = 0$. ◄—— **The two numbers are −1 and −2.**

$$(x - 1)(x - 2) = 0$$

KEY POINT **If two numbers are multiplied and the result is zero, then one or the other is zero.**

Either $x - 1 = 0$ or $x - 2 = 0$, giving the solution $x = 1$ or $x = 2$.

(b) Solve $x^2 - x - 56 = 0$. ◄—— **The two numbers are 7 and −8.**

$$(x + 7)(x - 8) = 0$$

Either $x + 7 = 0$ or $x - 8 = 0$, giving the solution $x = -7$ or $x = 8$.

Algebra

Solving quadratic equations by completing the square

AQA A   AQA B
EDEXCEL A   EDEXCEL B
OCR A   OCR B
OCR C
NICCEA
WJEC

This is a different approach. Look at this example.

> **Example**
>
> Solve $(x - 3)^2 = 25$
>
> $x - 3 = \pm 5$
>
> The solution is $x = 8$ or $x = -2$.

The only x term is inside the bracket to be squared. Taking the square root will leave just x and no x^2.

> Don't forget the negative value of the square root.

Any quadratic equation can be rearranged so that it can be solved this way.

It will help to remember that
$(x + p)^2 = x^2 + 2px + p^2$
and
$(x - q)^2 = x^2 - 2qx + q^2$

> **Examples**
>
> (a)   Solve
>
> $x^2 - 6x + 7 = 0$
>
> $x^2 - 6x = -7$
>
> $x^2 - 6x + 9 = -7 + 9$
>
> $(x - 3)^2 = 2$
>
> $x - 3 = \pm\sqrt{2}$

But $x^2 - 6x + 9 = (x - 3)^2$, so add 9 to each side.

> The solution is $x = 3 + \sqrt{2}$ or $x = 3 - \sqrt{2}$ (4.41 or 1.59 in decimals to 3 s.f.).
>
> (b)   Solve
>
> $2x^2 + 6x - 3 = 0$
>
> $x^2 + 3x = 1\frac{1}{2}$
>
> $x^2 + 3x + \left(1\frac{1}{2}\right)^2 = 1\frac{1}{2} + \left(1\frac{1}{2}\right)^2$
>
> $(x + 1.5)^2 = 3.75$
>
> $x + 1.5 = \pm\sqrt{3.75}$

Divide each side by 2.

$x^2 + 3x + \left(1\frac{1}{2}\right)^2 = \left(x + 1\frac{1}{2}\right)^2$, so add $\left(1\frac{1}{2}\right)^2$ to each side

> The solution is $x = 0.436$ or $x = -3.44$ to 3 s.f.

Solving quadratic equations by using the formula

AQA A   AQA B
EDEXCEL A   EDEXCEL B
OCR A   OCR B
OCR C
NICCEA
WJEC

If you apply the method of completing the square to the general equation $ax^2 + bx + c = 0$ you will obtain a formula for the solution of the equation. You do not have to find the formula for yourself and it can be quoted in an examination.

KEY POINT

The solution of the quadratic equation $ax^2 + bx + c = 0$ is given by
$$x = \frac{-b \pm \sqrt{b^2 - 4ac}}{2a}$$

Try the formula on the earlier examples.

(a) $x^2 - 6x + 7 = 0$

$$x = \frac{6 \pm \sqrt{36 - 4 \times 7}}{2}$$

Solution is $x = 4.41$ or $x = 1.59$.

(b) $2x^2 + 6x - 3 = 0$

Great care is needed with the negative signs and getting the operations on your calculator in the right order.

$$x = \frac{-6 \pm \sqrt{36 - 4 \times 2 \times (-3)}}{2 \times 2}$$

Solution is $x = 0.436$ or $x = -3.44$.

PROGRESS CHECK

1 Solve these equations by factorising.
   (a) $x^2 + 9x + 20 = 0$ (b) $x^2 - 4x - 45 = 0$ (c) $x^2 + 3x - 40 = 0$
   (d) $x^2 - 7x + 6 = 0$
2 Solve these equations by completing the square.
   (a) $y^2 - 8y + 3 = 0$ (b) $2y^2 + 10y - 9 = 0$
3 Solve these equations using the formula.
   (a) $x^2 + 3x + 1 = 0$ (b) $3x^2 - 8x - 7 = 0$

1 (a) $x = -4$ or $x = -5$ (b) $x = 9$ or $x = -5$ (c) $x = -8$ or $x = 5$ (d) $x = 6$ or $x = 1$
2 (a) $y = 7.61$ or $y = 0.394$ (b) $y = 0.779$ or $y = -5.78$
3 (a) $x = -0.382$ or $x = -2.62$ (b) $x = 3.36$ or $x = -0.694$

2.10 Simultaneous linear and quadratic equations

LEARNING SUMMARY

After studying this section, you will be able to:

● *solve simultaneous equations when one of them is quadratic*

Solving simultaneous equations when one equation is quadratic

AQA A   AQA B
EDEXCEL A   EDEXCEL B
OCR A   OCR B
OCR C
NICCEA
WJEC

Two methods were used to solve linear simultaneous equations. Only the substitution method will be used when one equation is quadratic.

KEY POINT

To solve simultaneous equations when one is quadratic, use one equation to substitute for one unknown in the other equation.

Example
Solve the simultaneous equations:

$$y = 2x^2$$

$$2y - x = 4$$

Either equation can be used for substitution. Use the first and substitute for y in the second.

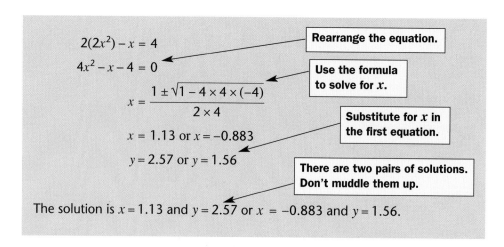

$$2(2x^2) - x = 4$$

Rearrange the equation.

$$4x^2 - x - 4 = 0$$

Use the formula to solve for x.

$$x = \frac{1 \pm \sqrt{1 - 4 \times 4 \times (-4)}}{2 \times 4}$$

$$x = 1.13 \text{ or } x = -0.883$$

Substitute for x in the first equation.

$$y = 2.57 \text{ or } y = 1.56$$

There are two pairs of solutions. Don't muddle them up.

The solution is $x = 1.13$ and $y = 2.57$ or $x = -0.883$ and $y = 1.56$.

Here is another example.

> **The circle has centre the origin and radius 5. You can use Pythagoras' theorem to show that it has the given equation (see page 77).**

Example

Find the coordinates of the points where the straight line $2y + x = 6$ meets the circle $x^2 + y^2 = 25$.

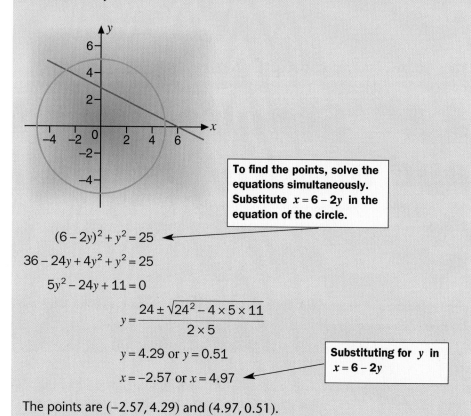

To find the points, solve the equations simultaneously. Substitute $x = 6 - 2y$ in the equation of the circle.

$$(6 - 2y)^2 + y^2 = 25$$

$$36 - 24y + 4y^2 + y^2 = 25$$

$$5y^2 - 24y + 11 = 0$$

$$y = \frac{24 \pm \sqrt{24^2 - 4 \times 5 \times 11}}{2 \times 5}$$

$$y = 4.29 \text{ or } y = 0.51$$

$$x = -2.57 \text{ or } x = 4.97$$

Substituting for y in $x = 6 - 2y$

> **Check your solution by seeing if the points satisfy the equation of the circle.**

The points are $(-2.57, 4.29)$ and $(4.97, 0.51)$.

1 Solve these simultaneous equations.
  (a) $x + y = 7$, $x^2 + y^2 = 49$ (b) $y = 2x^2 - 1$, $2x + 3y = 7$
2 Find the points of intersection of the curve $y = x^2 - 2x - 3$ and the straight line $y = 2x - 1$.

> **PROGRESS CHECK**

2 $x = 4.45$ and $y = 7.90$ or $x = -0.45$ and $y = -1.90$

1 (a) $x = 0$ and $y = 7$ or $x = 7$ and $y = 0$ (b) $(1.14, 1.58)$, $(-1.47, 3.31)$

2.11 Numerical methods

LEARNING SUMMARY

After studying this section, you will be able to:
- *solve equations that you cannot solve by simple manipulative methods*

Trial and improvement

AQA A   AQA B
EDEXCEL A   EDEXCEL B
OCR A   OCR B
OCR C
NICCEA
WJEC

Some equations cannot be solved in a simple way.
Here are two cubic equations.

$$(2x + 1)(x - 3)(x - 5) = 0$$

$$x^3 + 2x - 1 = 0$$

The first is in a factorised form and can be solved in a similar way to a quadratic equation. One of the factors, $(2x + 1)$, $(x - 3)$ and $(x - 5)$, must be zero so $x = -\frac{1}{2}$ or $x = 3$ or $x = 5$.

> **You will not be expected to find the factors of a cubic equation.**

If the cubic equation cannot be factorised there is no easy formula for the solution.

However, it is possible to use an approximate method, which can be refined to any degree of accuracy required.

For example, in the second equation, when $x = 0$ the left-hand side is -1, which is less than 0;

when $x = 1$ the left-hand side is 2, which is greater than 0.

> **KEY POINT**
>
> The object of the method of trial and improvement is to systematically find a value of x which makes the expression $x^3 + 2x - 1$ as close as possible to 0. This value is called a **root** of the equation.

$0.5^3 + 2 \times 0.5 - 1 = 0.125$, still too big.   ← | Try half-way between 0 and 1. |

$0.3^3 + 2 \times 0.3 - 1 = -0.373$, now too small.   ← | Now try between 0 and 0.5, say 0.3. |

> **Never use trial and improvement when a manipulation method will work.**

$0.4^3 + 2 \times 0.4 - 1 = -0.136$, still too small.   ← | Try 0.4. |

| Try half way between 0.4 and 0.5. |

$0.45^3 + 2 \times 0.45 - 1 = -0.008\ 875$, which is quite close.

$0.47^3 + 2 \times 0.47 - 1 = 0.043\ 823$ ← | Again, try about half-way, say 0.47. |

$0.46^3 + 2 \times 0.46 - 1 = 0.017\ 336$ ← | Now try 0.46. |

> **Greater accuracy can be achieved by continuing the process. The next value to try is between 0.45 and 0.455.**

$0.455^3 + 2 \times 0.455 - 1 = 0.004\ 196\ 375$ ← | Now try 0.455. |

The root of the equation between 0 and 1 is 0.45, to 2 d.p.   ←

| If the root was required correct to two decimal places, the result would be 0.45, not 0.46, as 0.455 is too big. |

PROGRESS CHECK

1 Use the method of trial and improvement to find a root for each of these equations, correct to one decimal place.
(a) $x^3 - 4x + 5 = 0$, between -3 and -2
(b) $x^3 + x^2 - 3 = 0$, between 1 and 2   (c) $x^4 + 3x - 6 = 0$, between -2 and -1.

1 (a) -2.4 (b) 1.2 (c) -1.8

2.12 Sequences

LEARNING SUMMARY

After studying this section, you will be able to:

- generate sequences of integers
- describe the nth term of a sequence

Generating sequences

> **KEY POINT**
> To generate a sequence, you need a **starting value** and a **rule** to find the next term.

For example,

Starting with **1**, add **2** each time.

This gives the sequence **1, 3, 5, 7, 9, ...**, which is the odd numbers.

Here is another **example**.

Start with **1** again and now multiply by **2** each time.

This gives the sequence **1, 2, 4, 8, 16, ...**, which is the powers of 2.

> **KEY POINT**
> Sequences can also be generated from an expression for a general term, the nth term.

Here n stands for the number of the position in the sequence, so in the first position, $n = 1$, in the second $n = 2$, and so on.

For example,

if the **nth term** of a sequence is **$2n - 1$**, then the sequence is **1, 3, 5, 7, 9, ...**, the odd numbers again.

Finding the nth term

Try some more sequences.

nth term $= n + 1$ gives 2, 3, 4, 5, 6, ...

nth term $= n - 3$ gives $-2, -1, 0, 1, 2, ...$

In each case the terms go up by one each time.

Now try these.

nth term $= 2n + 1$ gives 3, 5, 7, 9, ...
nth term $= 2n + 8$ gives 10, 12, 14, 16, ...

The difference is 2 each time.

nth term $= 5n - 4$ gives 1, 6, 11, 16, 21, ...

The differences are now 5.

> In algebra, a number which is the multiple of an unknown is called its **coefficient**. For example in the expression $2x^2 + 5x - 3$, the coefficient of x^2 is 2 and the coefficient of x is 5.

 KEY POINT

> If the differences between successive terms in a sequence are constant, this difference gives the coefficient of n in the expression for the nth term.

For example, to find the **nth term** of the sequence

 3, 6, 9, 12, 15, ...

The differences are 3 each time, so the **nth term** involves **3n**.
Check to see that this is in fact the nth term.

Here is another **example**.
Find the nth term of the sequence

 5, 9, 13, 17, 21, ...

The differences are 4 each time, so the **nth term** involves **4n**.
However, **4n** gives the sequence **4, 8, 12, 16,** ... which is always one less than the given sequence.
This means that the **nth** term is **4n + 1**.
Check to see that it works.

PROGRESS CHECK

1 Generate these sequences.
   (a) Start with −2 and add 5 each time.
   (b) Start with 50 and subtract 3 each time.
   (c) Start with 1 and multiply by 3 each time.
   (d) Start with 128 and multiply by $\frac{1}{2}$ each time.
2 Write down the first four terms of the sequences with these nth terms.
   (a) $4n - 3$ (b) $2n^2$ (c) $n(n+1)$ (d) 2^n
3 Find the nth terms for these sequences.
   (a) 1, 4, 7, 10, 13, ... (b) 4, 6, 8, 10, 12, ... (c) −5, 0, 5, 10, 15, ...
   (d) 26, 25, 24, 23, 22, ...

3 (a) $3n - 2$ (b) $2n + 2$ (c) $5n - 10$ (d) $27 - n$
2 (a) 1, 5, 9, 13, ... (b) 2, 8, 18, 32, ... (c) 2, 6, 12, 20, ... (d) 2, 4, 8, 16, ...
   (d) 128, 64, 32, 16, 8, ...
1 (a) −2, 3, 8, 13, 18, ... (b) 50, 47, 44, 41, 38, ... (c) 1, 3, 9, 27, 81, ...

 **Algebra**

2.13 Graphs of linear functions

LEARNING SUMMARY

After studying this section, you will be able to:

● *recognise the form of the equation for a straight line*
● *find and interpret the gradients of straight lines*

> The function is sometimes written as a mapping $f : x \rightarrow ax + b$.

KEY POINT

A function f of x is linear when $f(x) = ax + b$, where a and b are constants. The graph of the function $y = ax + b$ is a straight line.

Equations of straight lines

AQA A   AQA B
EDEXCEL A   EDEXCEL B
OCR A   OCR B
OCR C
NICCEA
WJEC

Draw the graph of $y = 3x + 1$.

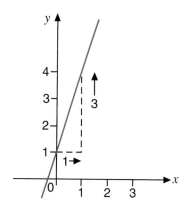

> The gradient is found by dividing the increase in y values by the increase in x values.

The graph goes up to the right. For every unit it moves across, it moves 3 units up. The **gradient** of the line is 3.

It crosses the y-axis at (0, 1). This point is called the **intercept**.
Notice where the 3 and the 1 appear in the equation.

Now draw the graph of $y + 2x - 3 = 0$. This can be written as $y = -2x + 3$.

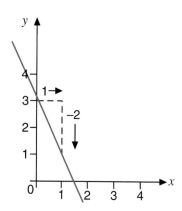

The gradient is negative since the *y* values decrease.

The graph goes down to the right. For every unit it moves across it moves 2 units down. The gradient of the line is −2.

The line crosses the *y*-axis at (0, 3).

Notice where the −2 and the 3 appear in the equation.

> **KEY POINT**
> In the equation of a straight line $y = mx + c$, m is the gradient and c is the intercept on the *y*-axis.

Examples

Find the gradients and *y*-intercepts for these straight-line equations.

(a) $y = x - 1$ (b) $x + y = 5$ (c) $2y = x - 3$

(a) the gradient and intercept can be read directly from the equation. The gradient is 1 and the intercept is −1.

Rewrite (b) in the form $y = mx + c$.

(b) $y = -x + 5$. Gradient is −1 and intercept is 5.

Divide each side of (c) by 2.

(c) $y = \frac{1}{2}x - 1\frac{1}{2}$. Gradient is $\frac{1}{2}$ and intercept is $-1\frac{1}{2}$.

Parallel and perpendicular lines

AQA A AQA B
EDEXCEL A EDEXCEL B
OCR A OCR B
OCR C
NICCEA
WJEC

Draw the graph of $y = 3x - 1$ on the same axes as $y = 3x + 1$.

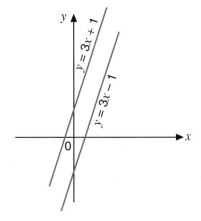

These lines are parallel. Try $y = 3x + 3$. Still parallel?

> **KEY POINT**
> Lines with equal gradients are parallel.

Now draw a straight line through (0, 1) perpendicular to the three lines.

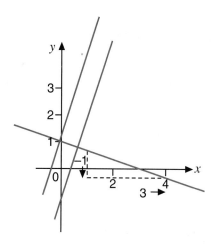

What is the gradient of this new line?

It goes down one unit for every 3 it moves across.

This gives gradient $-\frac{1}{3}$.

Multiply the gradients of the two perpendicular lines. Try this for more pairs of perpendicular lines.

> The gradient is negative as the y-values decrease.

KEY POINT

Multiplying the gradients of two perpendicular lines gives –1.

Examples

Find the gradient of a line perpendicular to $y = \frac{1}{2}x + 2$.

Let the gradient be m, then

$m \times \frac{1}{2} = -1$, giving $m = -2$.

PROGRESS CHECK

1 Find the gradients and y-intercepts of these lines.

(a) $y = -\frac{1}{2}x + 3$ (b) $y = 4 - 2x$ (c) $2x - y = 1$ (d) $3y = 2x - 1$

2 Find the equations of these lines.
(a) Parallel to $y = 4x - 2$, through $(0, 3)$
(b) Parallel to $x + y = 5$ through $(0, -2)$
(c) Perpendicular to $y = 2x + 5$, through $(0, 4)$
(d) Perpendicular to $x + y = 5$, through $(0, -1)$

2 (a) $y = 4x + 3$ (b) $y = -x - 2$ or $x + y = -2$ (c) $y = -\frac{1}{2}x + 4$ or $2y + x = 8$ (d) $y = x - 1$

1 (a) $-\frac{1}{2}$, 3   (b) -2, 4   (c) 2, -1 (d) $\frac{2}{3}$, $-\frac{1}{3}$

2.14 *Interpreting graphical information*

LEARNING SUMMARY

After studying this section, you will be able to:

● *use and interpret graphs involving practical situations*

Graphs of real-life situations

AQA A   AQA B
EDEXCEL A EDEXCEL B
OCR A   OCR B
OCR C
NICCEA
WJEC

The graphs in this section will show how quantities vary with time.

Examples

(a) This is a distance-time graph.

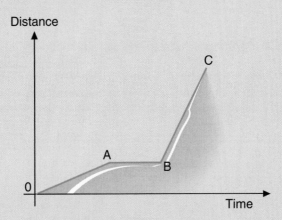

> The gradient of a distance-time graph gives the velocity (speed).

Look at each section.
From O to A, distance increases as time increases steadily. This is constant speed.
From A to B, the distance does not increase. The object is stationary.
From B to C, the speed is again constant but greater, as the gradient is greater.

(b) This is a velocity-time graph.

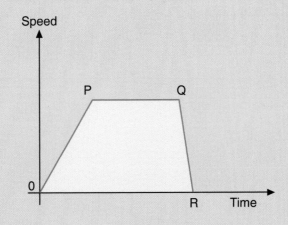

> The gradient of a velocity (speed) - time graph gives the acceleration.

Look at each section.

From O to P, the speed increases steadily. This is constant acceleration.

From P to Q, there is no increase in speed. There is no acceleration.

From Q to R, the speed decreases. This is negative acceleration or deceleration.

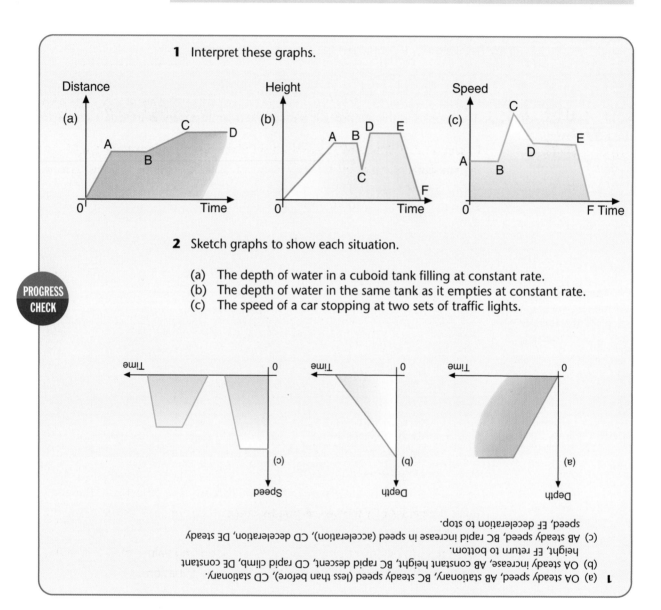

1 Interpret these graphs.

(a) Distance / Time

(b) Height / Time

(c) Speed / F Time

2 Sketch graphs to show each situation.

(a) The depth of water in a cuboid tank filling at constant rate.

(b) The depth of water in the same tank as it empties at constant rate.

(c) The speed of a car stopping at two sets of traffic lights.

1 (a) OA steady speed, AB stationary, BC steady speed (less than before), CD stationary.

(b) OA steady increase, AB constant height, BC rapid descent, CD rapid climb, DE constant height, EF return to bottom.

(c) AB steady speed, BC rapid increase in speed (acceleration), CD deceleration, DE steady speed, EF deceleration to stop.

2.15 *Quadratic and other functions*

> *After studying this section, you will be able to:*
> ● **solve equations using graphs**
> ● **draw the graphs of functions**

Graphical solutions

AQA A   AQA B
EDEXCEL A   EDEXCEL B
OCR A   OCR B
OCR C
NICCEA
WJEC

In section 2.10 (page 59) you solved simultaneous equations to find the points of intersections of the graphs. These points also give the solution to the equations.

For example, Look at these graphs.

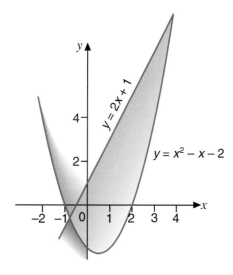

Read the values of x where the graphs meet.
$x = -0.8$, $x = 3.8$.
This time, do not substitute to find the corresponding values of y. The values of x are the roots of a quadratic equation. What is the equation?

$2x + 1 = x^2 - x - 2$ ⟵ | **Substitute $2x + 1$ for y.** |

You can check these using the formula.

$x^2 - 3x - 3 = 0$, which has roots -0.8 and 3.8, approximately.
⟵ | **Rearranging** |

Look again at this graph.

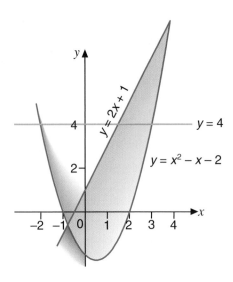

By drawing other straight lines, you can solve a range of equations.

Try $y = 0$ (the x-axis). The points of intersection will be the roots of $x^2 - x - 2 = 0$. These are -1 and 2.

Try $y = 4$. The points of intersection will be the roots of $x^2 - x - 2 = 4$, or $x^2 - x - 6 = 0$. These are -2 and 3.

Graphs of functions

AQA A   AQA B
EDEXCEL A   EDEXCEL B
OCR A   OCR B
OCR C
NICCEA
WJEC

There is a range of functions that you are expected to be able to graph. You should also be able to recognise their characteristic shapes.

> The use of unknown constants, a, b, c, etc may look complicated but in an examination they would be known numbers.

Here are some examples.

$y = ax^3 + bx^2 + cx + d$ (cubic); $y = \dfrac{1}{x}$ (reciprocal);

$y = k^x$ (exponential); $y = \sin x$; $y = \cos x$.

Examples

(a) Draw the graph of $y = x^3 - x + 2$, from $x = -2$ to $x = 2$.

> Work out a table of values.

> You need to find the values at -0.5 and 0.5 to find the right shape.

x	-2	-1	-0.5	0	0.5	1	2
y	-4	2	2.375	2	1.625	2	8

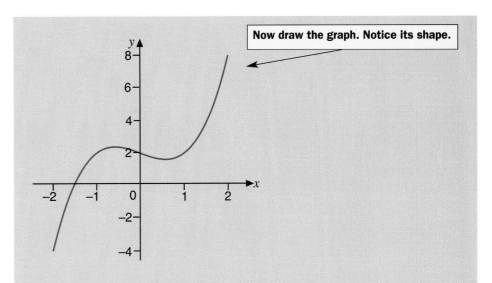

Now draw the graph. Notice its shape.

(b) Draw the graph of $y = \dfrac{2}{x}$ from $x = -4$ to $x = 4$, excluding 0.

Work out a table of values.

x	−4	−3	−2	−1	−0.5	0.5	1	2	3	4
y	−0.5	−0.67	−1	−2	−4	4	2	1	0.67	0.5

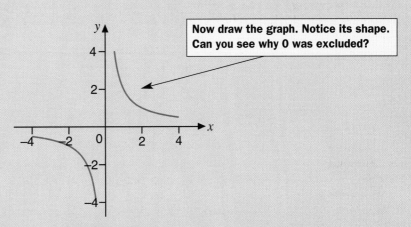

Now draw the graph. Notice its shape. Can you see why 0 was excluded?

(c) Draw the graph of $y = \sin x°$ from $x = 0$ to $x = 360$.

Your calculator will work out the values for you.

Work out a table of values.

There is no need to work out all the points as you can see that the negative values follow the positive ones.

x	0	30	60	90	120	150	180	210	270	360
y	0	0.5	0.87	1	0.87	0.5	0	−0.5	−1	0

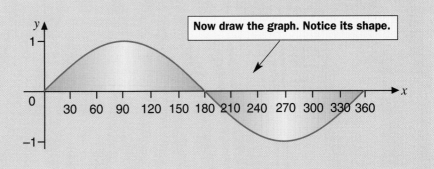

Now draw the graph. Notice its shape.

(d) Draw the graph of $y = 2^x$ from $x = -5$ to $x = 5$.

Work out a table of values.

Do not record too many figures in the table. You cannot plot 0.03125 that accurately!

x	−5	−4	−3	−2	−1	0	1	2	3	4	5
y	0.03	0.06	0.125	0.25	0.5	1	2	4	8	16	32

Now draw the graph. Notice its shape.

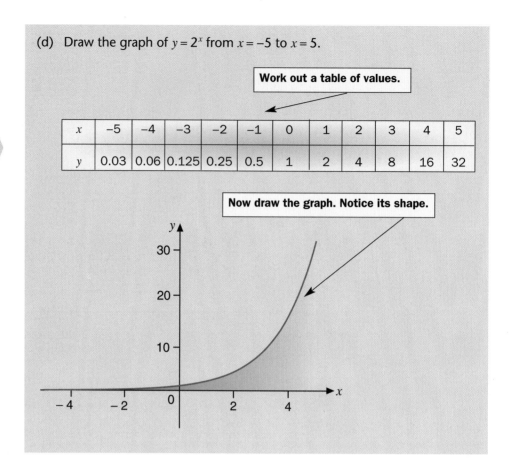

PROGRESS CHECK

1 Draw the graphs of these functions.

(a) $y = x^3 + 2x - 1 = 0$, $-2 \leqslant x \leqslant 2$ (b) $y = \cos x°$, $0 \leqslant x \leqslant 360$

(c) $y = \dfrac{3}{x}$, $-3 \leqslant x < 0$, $0 < x \leqslant 3$

2 Write down functions for which these could be the sketch graphs.

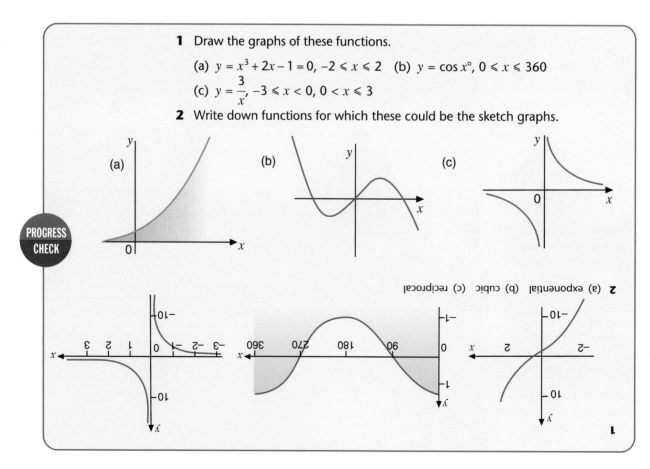

2 (a) exponential (b) cubic (c) reciprocal

1

2.16 Transformation of functions

After studying this section, you will be able to:

- apply transformations to the graphs of various functions

Transforming graphs

AQA A   AQA B
EDEXCEL A   EDEXCEL B
OCR A   OCR B
OCR C
NICCEA
WJEC

The sketch graphs show the functions $y = f(x)$ and $y = f(x) + k$.

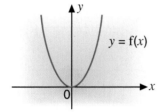 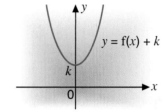

In relation to $y = f(x)$, where is the graph of $y = f(x) + k$, where k is a positive constant?

> **KEY POINT**
>
> For every value of x, y will be k more, that is the graph of $y = f(x)$ is translated by k in the y-direction, i.e., $\begin{pmatrix} 0 \\ k \end{pmatrix}$.

For example, the graph of $y = x^2 + 2$ is the graph of $y = x^2$ translated **2** units in the **y-direction**.

If in doubt, substitute a numerical value.

In relation to $y = f(x)$, where is the graph of $y = f(x + k)$, where k is a positive constant?

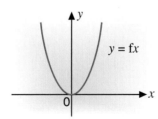 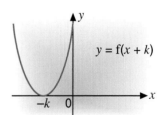

> **KEY POINT**
>
> For every value of x, y will take the value it would have had at $x + k$, that is the graph of $y = f(x)$ is translated by $-k$ in the x-direction, i.e., $\begin{pmatrix} -k \\ 0 \end{pmatrix}$.

For example, the graph of $y = \cos(x + 90)°$ is the graph of $y = \cos x°$ translated **90° to the left.**

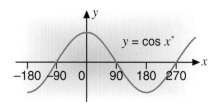

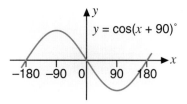

In relation to $y = f(x)$, where is the graph of $y = kf(x)$, where k is a positive constant?

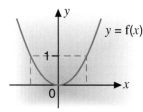

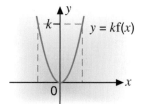

KEY POINT For every value of x, y will be k times bigger, that is the graph of $y = f(x)$ **is stretched by a factor of k in the y-direction.**

For example, the graph of $y = 4x^3$ is the graph of $y = x^3$ stretched by a factor of **4** in the **y-direction.**

> The effect is similar to the previous transformation but it is not the same.

In relation to $y = f(x)$, where is the graph of $y = f(kx)$, where k is a positive constant?

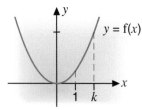

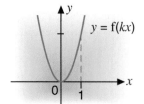

KEY POINT For every value of x, y will take the value it would have had at kx, that is the graph of $y = f(x)$ is stretched by a factor $\dfrac{1}{k}$ in the x-direction.

For example, the graph of $y = \sin 2x$ is the graph of $y = \sin x$ stretched by factor $\frac{1}{2}$ (reduced by a factor 2) in the x-direction.

> All the graphs in this section are sketch graphs. You should show the correct shape and indicate where they cross the axes but they need not be accurate.

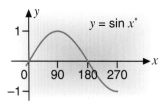

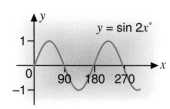

PROGRESS CHECK

1 Find the transformation in each case. Sketch the functions.
(a) $y = x + 2$, $y = x - 2$ (b) $y = x^2$, $y = (x - 3)^2$ (c) $y = \cos x$, $y = 2\cos x$
(d) $y = \sin x°$, $y = \sin(x - 180)°$

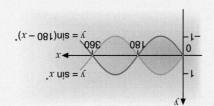
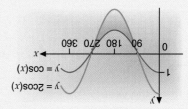
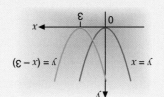
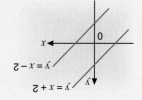

1 (a) Translation 4 down or $\begin{pmatrix} 0 \\ -4 \end{pmatrix}$ (b) Translation 3 to the right or $\begin{pmatrix} 3 \\ 0 \end{pmatrix}$
(c) Stretch factor 2 in y-direction (d) Translation $\begin{pmatrix} 180 \\ 0 \end{pmatrix}$

2.17 *Loci*

LEARNING SUMMARY

After studying this section, you will be able to:
- **find loci involving straight lines and circles**
- **find the equations of circles**

Constructing loci

AQA A   AQA B
EDEXCEL A   EDEXCEL B
OCR A   OCR B
OCR C
NICCEA
WJEC

These are the loci you should be able to construct.

(a) The locus of points equidistant from two fixed points A and B.

'Loci' is a Latin word and is the plural of 'locus'.

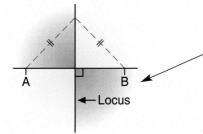

← Locus

The locus is formed by the vertices of isosceles triangles with base AB.

KEY POINT   The locus is the perpendicular bisector of the line AB.

Example

Draw the locus of points equidistant from (3, 2) and (7, 2).

What is its equation?

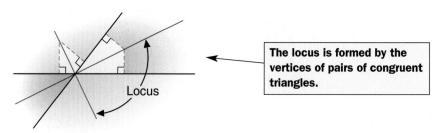

It is also the locus of points equidistant from (3, y) and (7, y), whatever the value of y.

The equation is $x = 5$.

(b)   The locus of points equidistant from two fixed lines.

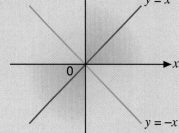

The locus is formed by the vertices of pairs of congruent triangles.

KEY POINT   The locus is the bisectors of the angles between the two lines.

Example

Draw the locus of points equidistant from both axes.

What is its equation?

$y = x$

$y = -x$

The equations are $y = x$ and $y = -x$.

Graphs of circles

AQA A   AQA B
EDEXCEL A   EDEXCEL B
OCR A   OCR B
OCR C
NICCEA
WJEC

There was some work on this in Section 2.10, page 60.

KEY POINT   A circle is the locus of points at a fixed distance from a fixed point.

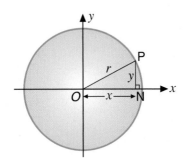

The radius of the circle is r, which is constant for a given circle.

Let the coordinates of a point P on the circle be (x, y).

$$x^2 + y^2 = r^2$$

> Apply Pythagoras' theorem to triangle OPN.

As P is any point on the circle, the relationship is true for all points on the circle and is therefore its equation.

> **KEY POINT**
> The equation of a circle, centre the origin and radius r, is $x^2 + y^2 = r^2$.

Example

Find the locus of points which are equidistant from the axes and 5 units from the origin.

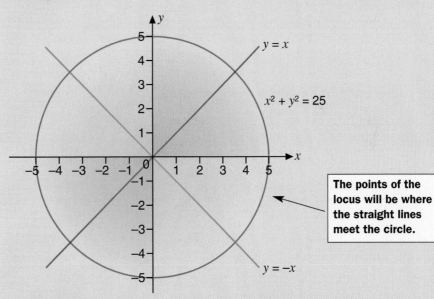

> The points of the locus will be where the straight lines meet the circle.

> The first two are the angle bisectors found earlier.

The equations are $y = x$, $y = -x$ and $x^2 + y^2 = 25$.

Find the intersection of $y = x$ and $x^2 + y^2 = 25$.
Substitute x for y.

$$x^2 + x^2 = 25$$
$$2x^2 = 25$$
$$x^2 = 12.5$$
$$x = \pm\sqrt{12.5} = \pm 3.54$$

> Since $y = x$, the values of y are also ± 3.54.

> Remember that the pairs of values go together in the same order, positive with positive and negative with negative.

Find the intersection of $y = -x$ and $x^2 + y^2 = 25$.
Substitute $-x$ for y.

Since $(-x)^2 = x^2$

$x^2 + x^2 = 25$

$2x^2 = 25$

Since $y = -x$, the values of y are now -3.54 or 3.54.

$x^2 = 12.5$

$x = \pm\sqrt{12.5} = 3.54$ or -3.54

The locus is four points $(3.54, 3.54)$, $(3.54, -3.54)$, $(-3.54, 3.54)$, $(-3.54, -3.54)$.

Remember that the pairs of values go together in the same order, positive with negative and negative with positive.

1 Find the equations of these loci.
(a) Points equidistant from $(1, 1)$ and $(1, 7)$
(b) Circle centre O and radius 6
2 Find all the points that are 6 units from O and equidistant from $(1, 1)$ and $(1, 7)$.

1 (a) $y = 4$ (b) $x^2 + y^2 = 36$ 2 $(4.47, 4)$, $(-4.47, 4)$

Sample GCSE questions

1 (a) Expand the brackets and simplify the expressions.

    (i) $3x(x-3)-4(3x-2)$

    (ii) $(2x+3)(x-2)$ **[5]**

(b) Factorise these expressions.

    (i) $4xy^2-2x^2y$

    (ii) $x^2-3x-28$ **[4]**

(c) Simplify these expressions.

    (i) $(3xy^3)^2$

    (ii) $\dfrac{2a^{-2}b^3}{4ab^{-1}}$ **[4]**

Expand the brackets, remembering that $-4\times-2=8$	*(a)* *(i)* $3x^2-9x-12x+8$ ✔✔
Collect like terms.	$3x^2-21x+8$ ✔
Expand the brackets. Collect like terms.	*(ii)* $2x^2+3x-4x-6$ ✔
	$2x^2-x-6$ ✔
The common factor is $2xy$.	*(b)* *(i)* $2xy(2y-x)$ ✔✔
Find two numbers with sum -3 and product -28.	*(ii)* $(x-7)(x+4)$ ✔✔
Remember that $(y^3)^2=y^{3\times2}=y^6$	*(c)* *(i)* $9x^2y^6$ ✔✔
Either form is acceptable.	*(ii)* $\frac{1}{2}a^{-3}b^4$ or $\dfrac{b^4}{2a^3}$ ✔✔

2 An aircraft flew across the Atlantic, a distance of 3000 miles. On the return flight, the average speed was 100 miles per hour faster.

The total journey time was 10 hours.

What was the average speed on the first part of the journey?   **[10]**

This is an example of a multi-step question. You have to decide on your method and how to start the question.	*Let the average speed on the first part be x mph.*
Remember that time is distance divided by speed.	*Time for the first part* $=\dfrac{3000}{x}$ ✔
	Time for second part $=\dfrac{3000}{x+100}$ ✔
	Total time is 10 hours, giving equation
Multiply each side by $x(x+100)$.	$\dfrac{3000}{x}+\dfrac{3000}{x+100}=10$ ✔
Expand the brackets and collect terms.	$3000(x+100)+3000x=10x(x+100)$ ✔

Sample GCSE questions

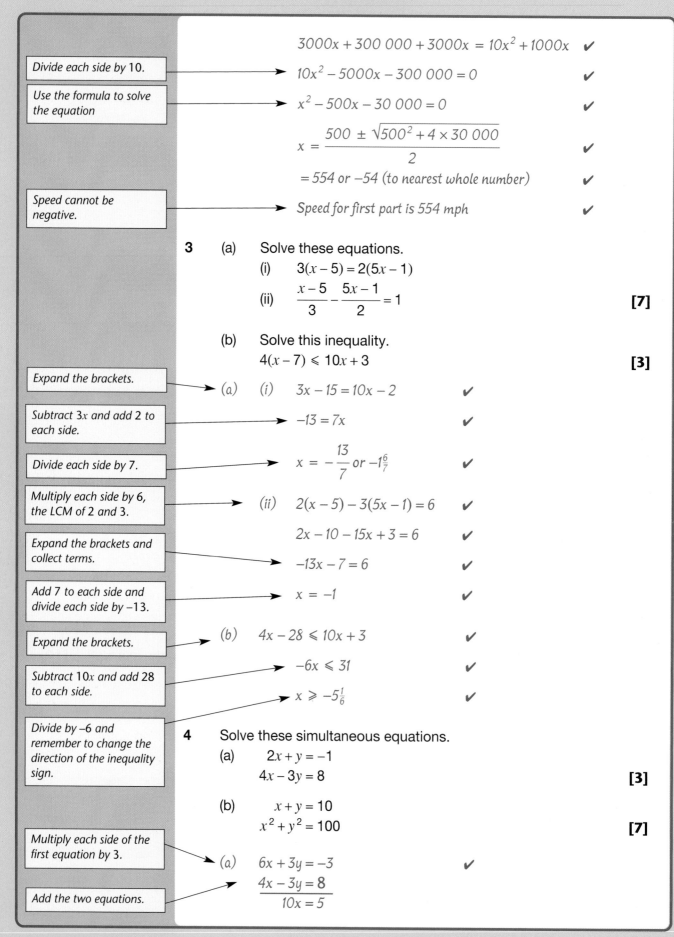

$$3000x + 300\,000 + 3000x = 10x^2 + 1000x \quad ✔$$

Divide each side by 10.

$$10x^2 - 5000x - 300\,000 = 0 \quad ✔$$

Use the formula to solve the equation

$$x^2 - 500x - 30\,000 = 0 \quad ✔$$

$$x = \frac{500 \pm \sqrt{500^2 + 4 \times 30\,000}}{2} \quad ✔$$

$$= 554 \text{ or } -54 \text{ (to nearest whole number)} \quad ✔$$

Speed cannot be negative.

Speed for first part is 554 mph ✔

3   (a)   Solve these equations.

     (i) $3(x - 5) = 2(5x - 1)$

     (ii) $\dfrac{x - 5}{3} - \dfrac{5x - 1}{2} = 1$ **[7]**

  (b)   Solve this inequality.

     $4(x - 7) \leqslant 10x + 3$ **[3]**

Expand the brackets.

(a)   (i) $3x - 15 = 10x - 2$ ✔

Subtract $3x$ and add 2 to each side.

$-13 = 7x$ ✔

Divide each side by 7.

$x = -\dfrac{13}{7} \text{ or } -1\frac{6}{7}$ ✔

Multiply each side by 6, the LCM of 2 and 3.

(ii) $2(x - 5) - 3(5x - 1) = 6$ ✔

Expand the brackets and collect terms.

$2x - 10 - 15x + 3 = 6$ ✔

$-13x - 7 = 6$ ✔

Add 7 to each side and divide each side by -13.

$x = -1$ ✔

Expand the brackets.

(b) $4x - 28 \leqslant 10x + 3$ ✔

Subtract $10x$ and add 28 to each side.

$-6x \leqslant 31$ ✔

$x \geqslant -5\frac{1}{6}$ ✔

Divide by -6 and remember to change the direction of the inequality sign.

4   Solve these simultaneous equations.

  (a) $2x + y = -1$

     $4x - 3y = 8$ **[3]**

  (b) $x + y = 10$

     $x^2 + y^2 = 100$ **[7]**

Multiply each side of the first equation by 3.

(a) $6x + 3y = -3$ ✔

Add the two equations.

$\dfrac{4x - 3y = 8}{10x = 5}$

Sample GCSE questions

$$x = \tfrac{1}{2} \qquad ✔$$

Substitute for x in the first equation.

$$2 \times \tfrac{1}{2} + y = -1$$

$$y = -2 \qquad ✔$$

(b) $$y = 10 - x \qquad ✔$$

Substitute for y in second equation.

$$x^2 + (10 - x)^2 = 100 \qquad ✔$$

$$x^2 + 100 - 20x + x^2 = 100$$

Divide each side by 2 and factorise.

$$2x^2 - 20x = 0$$

$$x^2 - 10x = 0 \qquad ✔$$

$$x(x - 10) = 0 \qquad ✔$$

Substitute for x in first equation

$$x = 0 \text{ or } x = 10 \qquad ✔$$

$$y = 10 \text{ or } y = 0 \qquad ✔$$

Take care to keep the pairs together.

Solution is $x = 0$ and $y = 10$, or $x = 10$ and $y = 0$. ✔

5 This formula is used in physics.

Notice that v occurs twice in the formula.

$$f = \frac{uv}{u + v}$$

Make v the subject of the formula. **[5]**

Multiply each side by $(u + v)$.

$$f(u + v) = uv \qquad ✔$$

Expand the brackets.

$$uf + vf = uv \qquad ✔$$

Change the sides over and subtract vf from each side.

$$uv - vf = uf \qquad ✔$$

$$v(u - f) = uf \qquad ✔$$

Factorise and divide each side by $(u - f)$.

$$v = \frac{uf}{u - f} \qquad ✔$$

6 (a) Draw the graph of $y = x^3 + 2x^2$ for $-3 \leqslant x \leqslant 2$. **[4]**

(b) Use your graph to solve $x^3 + 2x^2 = 4$. **[2]**

(c) Use trial and improvement to find this solution correct to two decimal places. **[4]**

(a)

x	−3	−2	−1.5	−1	−0.5	0	0.5	1	2
y	−9	0	1.125	1	0.375	0	0.625	3	16

✔✔

Find some values between the integer points to help draw the curve.

Sample GCSE questions

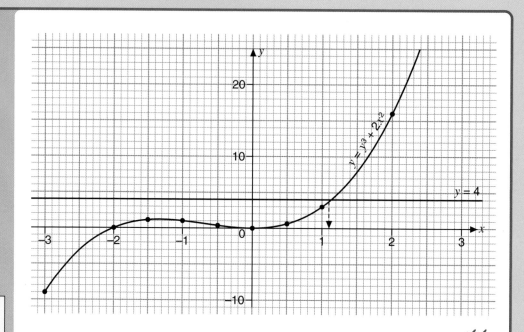

✔✔

Substituting y = in the equation of the graph gives the required equation.

(b) *Draw the line y = 4.*   ✔

  Read the graph, x = 1.1   ✔

You may have read this as 1.2. This is acceptable.

(c) $x = 1.2, x^3 + 2x^2 = 4.608$

  $x = 1.1, x^3 + 2x^2 = 3.751$ ✔

Too high, so the root is between 1.1 and 1.15.

  $x = 1.15, x^3 + 2x^2 = 4.16 ...$ ✔

This is very close, but too low.

  $x = 1.13, x^3 + 2x^2 = 3.996 ...$ ✔

This shows that the root is between 1.13 and 1.135 and is therefore 1.13 to 2 d.p.

  $x = 1.135, x^3 + 2x^2 = 4.0385 ...$ ✔

  Solution x = 1.13

The last mark is awarded only if the reason for selecting 1.13 is shown.

7 (a) Find the gradient of the straight line $2x + 5y = 15$. **[2]**

  (b) Find the equations of the straight lines through $(0, 2)$ that are parallel and perpendicular to the line in part (a). **[4]**

Rearrange in the form y = mx + c.

(a) $5y = 15 - 2x$

  $y = 3 - \frac{2}{5}x$ ✔

  $\text{Gradient} = -\frac{2}{5}$ ✔

There is 1 mark for the gradient and 1 for the intercept.

(b) *Parallel line:* $y = 2 - \frac{2}{5}x$ *or* $5y = 10 - 2x$ ✔✔

  Perpendicular line: $y = 2 + \frac{5}{2}x$ *or* $2y = 5x + 4$ ✔✔

Sample GCSE questions

8 The sketch shows the graph of $y = f(x)$.

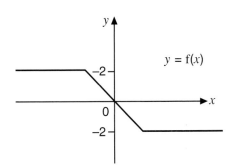

Draw sketches to show the graphs of

(a) $y = f(x) + 3$ (b) $y = f(x + 2)$ (c) $y = \frac{1}{2} f(x)$ **[5]**

In (a) the translation is $\begin{pmatrix} 0 \\ 3 \end{pmatrix}$, in (b) $\begin{pmatrix} -2 \\ 0 \end{pmatrix}$ and in (c) there is a stretch (reduction) of factor $\frac{1}{2}$. You should show this by marking the axes.

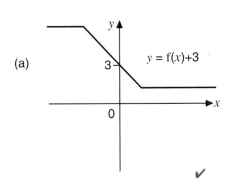

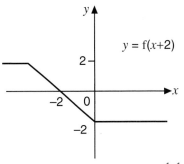

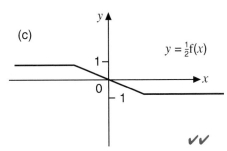

Exam practice questions

1 (a) Solve these equations.

      (i) $3(2 - x) = 2(x - 6)$

      (ii) $\dfrac{x}{2} - \dfrac{x - 1}{3} = 1$ **[6]**

   (b) Solve these inequalities.

      (i) $4x \leqslant 3(2x - 3)$

      (ii) $x^2 \leqslant 225$ **[5]**

2 (a) Make v the subject of the formula $\dfrac{1}{u} - \dfrac{1}{v} = \dfrac{1}{f}$. **[3]**

   (b) Make q the subject of the formula $p = \dfrac{qr}{r - q}$. **[4]**

3 Mark and Brian ordered cups of coffee and biscuits from a buffet bar.
Mark bought 2 cups of coffee and 3 biscuits. They cost £3.23.
Brian bought 3 cups of coffee and 1 biscuit. They cost £2.92.
Write down two simultaneous equations and solve them to find the cost of a cup of coffee
and the cost of a biscuit. **[5]**

4 (a) Draw the graph of $y = x^3 - 2x^2 + 1$ between $x = -1$ and $x = 3$. **[3]**

   (b) Use your graph to solve these equations.

      (i) $x^3 - 2x^2 + 1 = 0$

      (ii) $x^3 - 2x^2 + 1 = 5$ **[3]**

   (c) Use trial and improvement to find the solution of $x^3 - 2x^2 - 4 = 0$, correct to two
decimal places. **[4]**

5 Find the points of intersection of the straight line $2x + y = 3$ with the circle $x^2 + y^2 = 16$.
[7]

6 (a) Factorise $x^2 - 2x - 15$. **[2]**

   (b) Solve the equation $x^2 - 2x - 15 = 0$. **[1]**

   (c) Solve the inequality $x^2 - 2x - 15 > 0$. **[4]**

7 The sketch shows the graph of $y = \mathrm{f}(x)$.

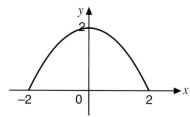

Sketch the graphs of

(a) $y = \mathrm{f}(x) + 1$ **[1]**

(b) $y = \mathrm{f}(x - 2)$ **[1]**

(c) $y = \dfrac{1}{2}\mathrm{f}(x)$ **[1]**

Shape and Space

Overview

Topic	Section	Studied in class	Revised	Practice questions
3.1 Congruent triangles	Tests for congruency			
3.2 Properties of triangles and quadrilaterals	Sum of angles of a triangle			
	Properties of quadrilaterals			
3.3 Right-angled triangles	Pythagoras' theorem			
	The length of the line joining two points on a graph			
	Problems in three dimensions			
3.4 Trigonometry	Trigonometric ratios			
	Area of a triangle			
	Sine, cosine and tangent of any angle			
	Sine and cosine rules			
3.5 Circles	Cyclic quadrilaterals			
	Angle theorems			
	Chord theorems			
	Tangents			
3.6 3-D shapes	Prisms			
3.7 Transformations	Reflections and translations			
	Rotations			
	Enlargements			
3.8 Area and volume	Dimensions			
	Areas and volumes of similar figures			
3.9 Vectors	Zero vector and unit vectors			
	Inverse vectors			
	Vector addition and subtraction			
3.10 Measures and constructions	Upper and lower bounds			
	Compound units			
	Constructions			
3.11 Mensuration	The sphere			
	Arcs, sectors and segments			
3.12 Loci	Three important loci			

3.1 Congruent triangles

Tests for congruency

KEY POINT

If two triangles are congruent they must have the same shape and the same size. This means that they will fit exactly onto each other when one of them is rotated, reflected or translated.

Two triangles are congruent if any of these conditions are satisfied:

● two sides and the included angle of one triangle are equal to the two sides and the included angle of the other triangle (side, angle, side or SAS)

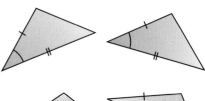

● the three sides of one triangle are equal to the corresponding three sides of the other triangle (side, side, side or SSS)

● two angles and a side in one triangle are equal to two angles and the corresponding side in the other triangle (angle, corresponding angle or AAS)

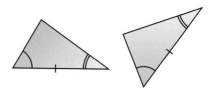

● each triangle is right-angled and the hypotenuse and one side of one triangle is equal to the hypotenuse and a side of the other triangle, (right angle, hypotenuse, side or RHS).

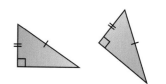

> The hypotenuse is the side opposite the right angle.

The tests for congruency: SSS; SAS; ASA are essentially the methods used to construct triangles (note that construction is discussed further in section 3.10):

(a) SSS

   (i) Draw the line AB of given length.

   (ii) Use compasses to construct arcs AC and BC.

   (iii) Draw AC and BC.

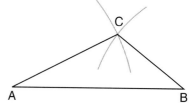

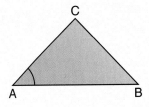
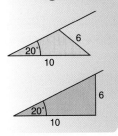
(b) SAS
   (i)   Draw a line AB of given length.
   (ii) Measure and draw the angle at A.
   (iii) Draw the line AC of given length.
   (iv) Join C to B.

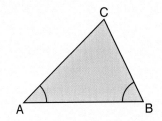

(c) ASA
   (i)   Draw the line AB of given length.
   (ii) Measure and draw the angles at A and B.
   (iii) Draw lines AC and BC.

The basic checks for congruency are:

(a) the three sides of each triangle are equal

(b) two sides and the included angle in each triangle are equal

(c) two angles and the corresponding side in each triangle are equal

(d) for a right-angled triangle the check is 'right angle, hypotenuse and side' in each triangle are equal.

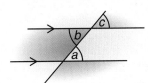

angle a = angle b (alternate angles)
angle a = angle c (corresponding angles)

PROGRESS CHECK

1 Which of these pairs of triangles are congruent?
Give a reason or explanation for your answer.

(a)

(b)

(c)

(d)

(e)

(f)

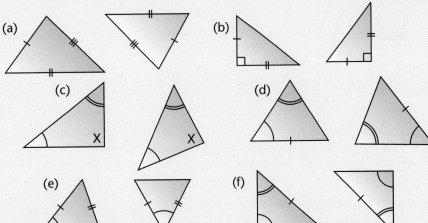

2 Triangle XYZ is isosceles with XY = XZ. The bisector of angle Y meets XZ at M; the bisector of angle Z meets XY at N. Prove that YM = ZN.

2
XY = XZ   ∴   angles XYZ and XZY are equal.
∴   because YM and ZN bisect the angles then angles MYZ and NZY are equal.
In triangles YMZ and ZNY, YZ is common.
∴   triangles YMZ and ZNY are congruent, (ASA)
∴   YM = ZN

1   Congruent pairs are: (a) (SSS); (b) (SAS); (d) and (f) (AAS) because the equal sides are in corresponding positions

3.2 Properties of triangles and quadrilaterals

After studying this section, you will be able to:

- *use the angle sum of a triangle and a quadrilateral*
- *identify quadrilaterals by their geometric properties*

Sum of angles of a triangle

You need to be able to prove that:

(a) the sum of the angles in a triangle is 180°.

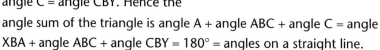

Take any triangle ABC. Construct XY through B and parallel to AC. Using the properties of parallel lines angle A = angle XBA and angle C = angle CBY. Hence the angle sum of the triangle is angle A + angle ABC + angle C = angle XBA + angle ABC + angle CBY = 180° = angles on a straight line.

You must remember the basic angle facts such as the sum of the angles on a straight line is 180°, and the properties of alternate and corresponding angles.

(b) the exterior angle of a triangle is equal to the sum of the interior opposite angles.

Take any triangle ABC. Construct a line through C, parallel to AB.

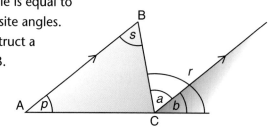

angle p = angle b
(corresponding angles)

angle s = angle a
(alternate angles)

∴ angle p + angle s = angle a + angle b

but angle r = angle a + angle b

∴ angle p + angle s = angle r

Properties of quadrilaterals

Sum of angles in a quadrilateral

You can use the fact that the sum of the angles in a triangle = 180° to prove that the angle sum of a quadrilateral is 360°.

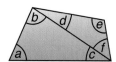

angles $a + b + c = 180°$

angles $d + e + f = 180°$

∴ $a + b + c + d + e + f = 360°$

Geometric properties of quadrilaterals

You need to be able to identify quadrilaterals by their geometric properties.

(a) Square

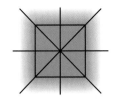

- all sides equal and opposite sides parallel
- all angles 90°
- four lines of symmetry
- rotational symmetry order 4
- diagonals bisect at right angles

(b) Rectangle

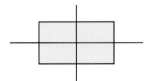

- opposite sides equal and parallel
- all angles 90°
- two lines of symmetry
- rotational symmetry order 2

(c) Parallelogram

- opposite sides equal and parallel
- opposite angles equal
- no lines of symmetry
- rotational symmetry order 2

(d) Rhombus

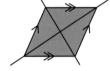

- all sides equal
- opposite sides parallel
- two lines of symmetry
- rotational symmetry order 2
- diagonals bisect at right angles

(e) Kite

- one line of symmetry
- diagonals intersect at right angles
- two pairs of adjacent sides equal

(f) Trapezium

- one pair of sides parallel

Note: isosceles trapezium has one line of symmetry

Examples

Find the size of the angles marked with letters:

(a)

(b)

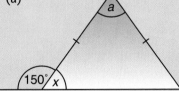

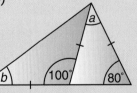

(a) Angle $x = 30°$ (angles on a straight line)
    Angle $a = 180° - 30° - 30° = 120°$ (angle sum of isosceles triangle)

(b) Angle $b = (180° - 100°) \div 2 = 40°$ (angle sum of isosceles triangle)
    Angle $a = 180° - 80° - 80° = 20°$ (angle sum of isosceles triangle)

PROGRESS CHECK

1 Find the size of the angles marked with letters:

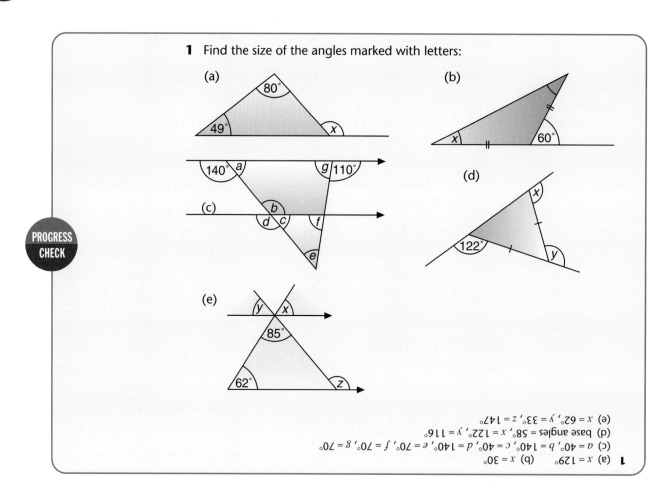

(a)

(b)

(c)

(d)

(e)

1 (a) $x = 129°$ (b) $x = 30°$
(c) $a = 40°$, $b = 140°$, $c = 40°$, $d = 140°$, $e = 70°$, $f = 70°$, $g = 70°$
(d) base angles = $58°$, $x = 122°$, $y = 116°$
(e) $x = 62°$, $y = 33°$, $z = 147°$

3.3 Right-angled triangles

LEARNING SUMMARY

After studying this section, you will be able to:

● **find the third side in a right-angled triangle**
● **find the length of a line joining two points on a graph**
● **use Pythagoras' theorem to solve problems in three dimensions**

Pythagoras' theorem

AQA
EDEXCEL
OCR
WJEC
NICCEA

KEY POINT

Pythagoras' theorem can be stated as:

In a right-angled triangle the area of the square on the hypotenuse = the sum of the areas of the squares on the other two sides.

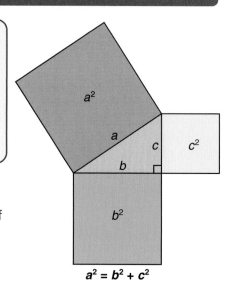

$$a^2 = b^2 + c^2$$

but it is normally abbreviated to:

The square on the hypotenuse = the sum of the squares on the other two sides.

Examples

Find the missing side in each of these triangles:

Always label the sides of the triangle, *a, b, c,* and then substitute in the formula before rearranging.

(a)

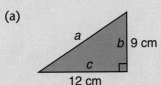

(b)

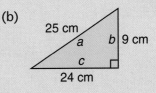

(a)　$a^2 = b^2 + c^2$
　　　$a^2 = 9^2 + 12^2$
　　　　$= 81 + 144$
　　　　$= 225$
　　　$a = \sqrt{225}$
　　　　$= 15$ cm

(b)　$a^2 = b^2 + c^2$
　　　$25^2 = b^2 + 24^2$
　　　$625 = b^2 + 576$
　　　　$b^2 = 625 - 576$
　　　　$b = \sqrt{49}$
　　　　　$= 7$ cm

It is a good idea to draw a sketch if a diagram isn't given. Try to draw it roughly to scale and mark on it any lengths you know. It may help you see any errors in your working.

 KEY POINT Learn Pythagoras' theorem: $a^2 = b^2 + c^2$

The length of the line joining two points on a graph

AQA A　AQA B
EDEXCEL A　EDEXCEL B
OCR A　OCR B
OCR C
NICCEA
WJEC

You can calculate the length of the line joining two points using Pythagoras' theorem.

Example

In the diagram A is at $(5, 5)$ and B is at $(18, 20)$.
Find the length AB.

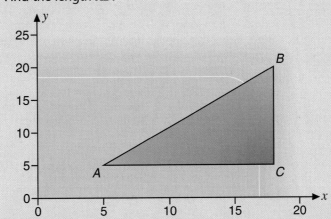

Find the lengths of AC and BC from the coordinates of A and B.

AC is the difference between the x-coordinates = 13
BC is the difference between the y-coordinates = 15

$AB^2 = AC^2 + BC^2$

　　　$= 13^2 + 15^2$

　　　$= 169 + 225$

　　　$= 394$

$AB = \sqrt{394}$

　　　$= 19.8$ to 3 s.f.

Problems in three dimensions

AQA A   AQA B
EDEXCEL A   EDEXCEL B
OCR A   OCR B
OCR C
NICCEA
WJEC

Example

The diagram shows a cuboid 5 cm by 4 cm by 12 cm.

Find the lengths of AC and AG.

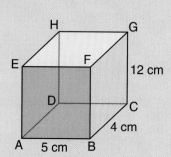

In triangle ABC: $AC^2 = 5^2 + 4^2$

$= 25 + 16$

$= 41$

$AC = 6.4$ cm

In triangle ACG: $AG^2 = AC^2 + CG^2$

$= 41 + 144$

$= 185$

$AG = 13.6$ cm

PROGRESS CHECK

1 Find the missing side in each triangle:

(a)

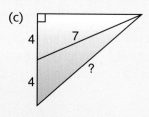

(b)   ?

(c)

2 A rectangular field has a length of 250 m and a width of 130 m. Find the length of the diagonal path across it.

3 A ladder, 6 m long, is resting against a wall. If the top of the ladder is 4 m above the ground, how far from the wall is the base of the ladder?

4 The diagram shows the side view of a shed.

Find the length of the sloping roof.

2.4 m   1.8 m

3.5 m

5 The diagonal of a rectangle is 2 cm longer than the length. If the width of the rectangle is 10 cm find the length.

6 Find the length of the diagonal of a rectangular room of length 4.8 m, width 3 m and height 2.6 m.

7 Find the distance between the following pairs of points:

(a) A(3, 4) and B(10, 6)   (b) A(4, 5) and B(−3, −8)

8 ABCD is a rectangular display area. AM is a vertical mast at the corner of the area. The top of the mast, M, is connected by straight supporting wires to the corners B, C and D.

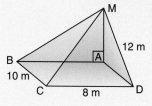

Calculate the height of the mast, AM, and the lengths of the wires, BM and CM.

8 AM = 6.6 m, BM = 10.4 m, CM = 14.4 m

7 (a) 7.28 (b) 14.76

6 6.2 m

5 $(x+2)^2 = x^2 + 10^2$ ∴ $x^2 + 4x + 4 = x^2 + 100$

$4x = 96$, ∴ $x = 24$ cm

4 3.6 m

1 (a) 6.7 (b) 9.4 (c) 9.8 **2** 281.8 m **3** 4.5 m

3.4 Trigonometry

After studying this section, you will be able to:

● use the sine, cosine and tangent ratios to find angles and sides in right-angled triangles

● find the area of a triangle using the formula $A = \dfrac{1}{2} ab \sin C$

● find the sine, cosine and tangent of any angle

● use the sine and cosine rules

Trigonometric ratios

You can use Pythagoras' theorem to solve problems in three dimensions: In a right-angled triangle the sides and the angles are related by three trigonometrical ratios: the sine (abbreviated to sin), the cosine, (abbreviated to cos) and the tangent (abbreviated to tan).

To use these remember that you need to identify which side is the hypotenuse (the longest side), which is the opposite (opposite the given angle), and which is the adjacent (next to the given angle), thus:

Label the sides hypotenuse, opposite and adjacent.

KEY POINT

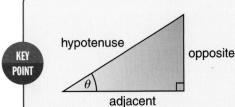

$$\sin \theta = \frac{\text{opposite}}{\text{hypotenuse}},$$

$$\cos \theta = \frac{\text{adjacent}}{\text{hypotenuse}},$$

$$\tan \theta = \frac{\text{opposite}}{\text{adjacent}}$$

You must remember these three ratios.

Examples

(a) Find the length, to 2d.p., of the side marked x in this triangle.

You have been given the length of the hypotenuse and need to find the opposite side, so you need to use the sine.

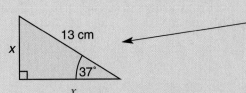

$$\sin 37 = \frac{x}{13} \text{ thus } x = 13 \times \sin 37$$

$$= 13 \times 0.6018 = 7.8235 = 7.82 \text{ cm to 2d.p.}$$

(b) Find the angle θ in this triangle:

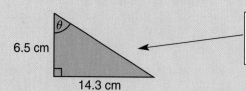

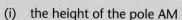

6.5 cm

14.3 cm

You are given the lengths of the opposite and adjacent so you need to use the tangent.

$$\tan \theta = \frac{\text{opposite}}{\text{adjacent}}$$

$$= \frac{14.3}{6.5}$$

$$= 2.2$$

$$\therefore \quad \theta = \tan^{-1} 2.2$$

$$= 65.6°$$

The 'tan^{-1}' shows that we need the angle whose tangent is 2.2.

(c) ABCD is a rectangular field. AM is a vertical pole at the corner A.
BC = 5 m, CD = 10 m and the angle of elevation of M from B is 25°.

Calculate
(i) the height of the pole AM
(ii) the angle of elevation of M from C.

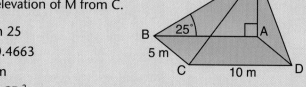

M

B

25°

A

5 m

C

10 m

D

(i) $AM = AB \tan 25$

$$= 10 \times 0.4663$$

$$= 4.66 \text{ m}$$

(ii) $AC^2 = AD^2 + CD^2$

$$= 25 + 100$$

$$AC = \sqrt{125} = 11.18 \text{ m}$$

$$\text{Tangent of angle } MCA = \frac{AM}{AC}$$

$$= \frac{4.66}{11.18} = 0.4168$$

$$\therefore \qquad \text{Angle } MCA = \tan^{-1} 0.4168$$

$$= 22.6°$$

Area of a triangle

AQA A   AQA B
EDEXCEL A   EDEXCEL B
OCR A   OCR B
OCR C
NICCEA
WJEC

You can calculate the area of a triangle using the sine of an angle.
The formula is:

KEY POINT

Area $= \dfrac{1}{2} ab \sin C$, or $A = \dfrac{1}{2} bc \sin A$, or $A = \dfrac{1}{2} ac \sin B$

Example

Find the area of this triangle.

$$\text{Area} = \frac{1}{2} ab \sin C$$

$$= \frac{1}{2} \times 14 \times 12 \times \sin 65$$

$$= 76.13 \text{ cm}^2$$

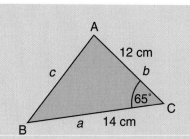

A

12 cm

c

b

65°

C

B

a

14 cm

Sine, cosine and tangent of any angle

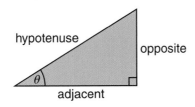

You already know that, in a right-angled triangle, the sine, cosine and tangent are defined as:

$$\sin \theta = \frac{\text{opposite}}{\text{hypotenuse}},$$

$$\cos \theta = \frac{\text{adjacent}}{\text{hypotenuse}},$$

$$\tan \theta = \frac{\text{opposite}}{\text{adjacent}}$$

You should also know that it is possible to define the trigonometric ratios for angles of any size using coordinates.

Draw a circle with radius 1 unit. The point P with coordinates (x, y) moves round the circumference of the circle.

OP makes an angle θ with the positive x-axis. The angle increases as P rotates anticlockwise.

For any angle θ the sine, cosine and tangent are given by the coordinates of P.

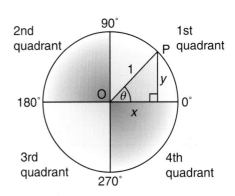

$$\sin \theta = \frac{y}{1} = y, \ \cos \theta = \frac{x}{1} = x, \ \tan \theta = \frac{y}{x}$$

Notice that as P rotates the coordinates change sign.
In the first quadrant, from 0° to 90° both y and x are positive.
In the second quadrant from 90° to 180° y is positive but x is negative.
In the third quadrant, from 180° to 270° both y and x are negative.
In the fourth quadrant, from 270° to 360° y is negative but x is positive.
Thus the sign of the sine, cosine and tangent of an angle changes according to

After 360° the pattern continues: the sine, cosine and tangent taking the appropriate sign according to the quadrant.

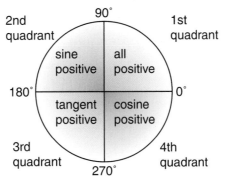

the size of the angle, that is the sign depends on which quadrant the angle is in.

The graphs of the three ratios, sine, cosine and tangent are shown on page 96 in Figures 1, 2 and 3. You need to recognise them and distinguish between them.

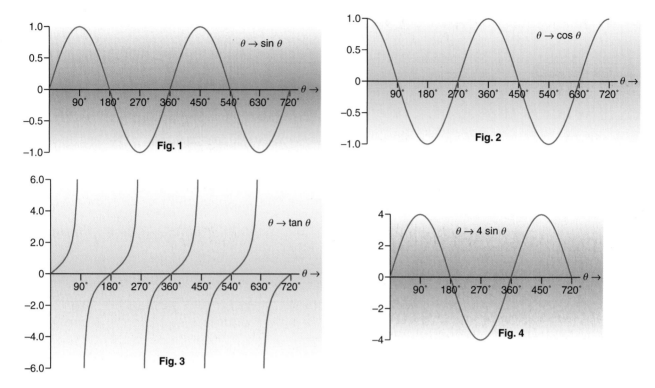

Fig. 1 · Fig. 2 · Fig. 3 · Fig. 4

The sine curve models many natural phenomena including sound and light waves. Because it repeats every 360° the graph is described as periodic with period 360°. (The other curves are similarly periodic with periods 360° for the cosine and 180° for the tangent.) Note that the tangent curve is of a different form and at 90°, 270°, 450° etc. the value of the tangent is infinity and hence cannot be plotted.

The graph of sin θ oscillates between −1 and +1. This means that the amplitude of the sine curve is 1. The amplitude can be changed by multiplying sin θ by a number, for example 4. The graph of 4 sin θ is shown in Figure 4.

You can use graphs like these to solve equations involving trigonometrical functions.

Example
Draw the graphs of $y = 6 \sin x$ and $y = 4$. Use these graphs to solve the equation $3 \sin x = 2$ for $0° \leqslant x \leqslant 360°$.

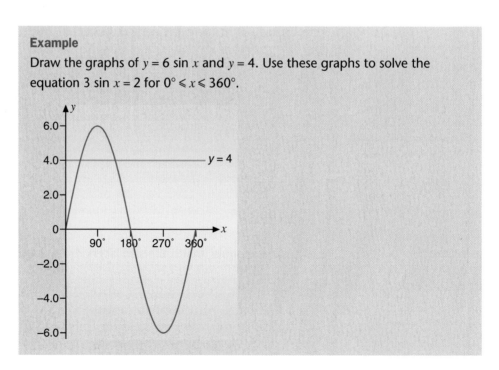

At the point where the graphs intersect $6 \sin x = 4$.
Dividing by 2 gives $3 \sin x = 2$.
The graphs intersect at approximately 40° and 135°.
(using a calculator gives 41.8° and 131.8°)

Sine and cosine rules

The sine and cosine rules apply to all triangles.

The sine rule

The sine rule is used to find the missing sides and angles in a triangle if: (a) the length of one side and the sizes of two angles are known, or (b) the lengths of two sides and the size of the angle opposite one of those two sides are known.

Remember that angles are shown with capital letters and sides with lower case letters.

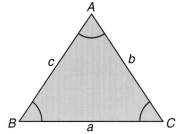

KEY POINT

The sine rule:
$$\frac{a}{\sin A} = \frac{b}{\sin B} = \frac{c}{\sin C} \quad \text{or} \quad \frac{\sin A}{a} = \frac{\sin B}{b} = \frac{\sin C}{c}$$

Examples

(a) In triangle *ABC*, find the missing sides and angle.

From the sine rule:

$$\frac{a}{\sin A} = \frac{b}{\sin B} = \frac{c}{\sin C}$$

$$\frac{9.5}{\sin 55} = \frac{c}{\sin 65}$$

$$c = \frac{9.5 \times \sin 65}{\sin 55} = 10.5 \text{ cm}$$

Angle $B = 180° - 65° - 55° = 60°$

$$\frac{9.5}{\sin 55} = \frac{b}{\sin 60}$$

$$b = \frac{9.5 \times \sin 60}{\sin 55} = 10.0 \text{ cm}$$

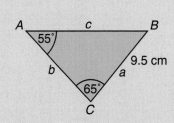

(b) Find the size of angle A in this triangle.

$$\frac{\sin A}{a} = \frac{\sin C}{c}$$

$$\frac{\sin A}{7} = \frac{\sin 40}{12}$$

$$\sin A = \frac{7 \times \sin 40}{12} = 0.375$$

$$\therefore \quad A = \sin^{-1} 0.375$$

$$= 22°$$

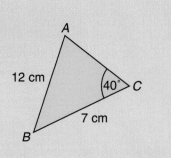

The cosine rule

The cosine rule: or $b^2 = c^2 + a^2 - 2ca \cos B$;
$a^2 = b^2 + c^2 - 2bc \cos A$ or $c^2 = a^2 + b^2 - 2ab \cos C$

There are two ways of using the cosine rule:

(a) to find a side when you are given two sides and the angle between them.

or $\cos B = \dfrac{c^2 + a^2 - b^2}{2ca}$

or $\cos C = \dfrac{a^2 + b^2 - c^2}{2ab}$

(b) to find an angle when you know the lengths of the three sides. In this case the cosine rule is used in this form:

$$\cos A = \frac{b^2 + c^2 - a^2}{2bc}$$

Remember to draw a realistic sketch.

Example

(a) In triangle ABC, $c = 10$ cm, $a = 12$ cm, angle $B = 20°$. Calculate the length of side b.

$$b^2 = c^2 + a^2 - 2ca \cos B$$

$$b^2 = 12^2 + 10^2 - 2 \times 12 \times 10 \times \cos 20$$

$$b = \sqrt{144 + 100 - 240 \cos 20}$$

$$= 4.3 \text{ cm}$$

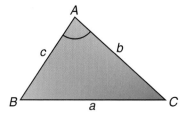

(b) Find the angles of a triangle with sides 4 cm, 5 cm and 6 cm.

$$\cos A = \frac{b^2 + c^2 - a^2}{2bc}$$

$$= \frac{4^2 + 5^2 - 6^2}{2 \times 4 \times 5}$$

$$= 0.125$$

$$\therefore \quad A = \cos^{-1} 0.125$$

$$= 82.8°$$

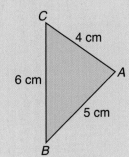

Using the sine rule to find angle B:

$$\frac{\sin B}{4} = \frac{\sin 82.8}{6}$$

$$\sin B = \frac{4 \times \sin 82.8}{6}$$

$$= 0.6614$$

$$\therefore \quad B = \sin^{-1} 0.6614$$

$$= 41.4°$$

$$\therefore \quad C = 180° - 41.4° - 82.8° = 55.8°$$

1 Find the missing angles and sides for each of these triangles:

(a)

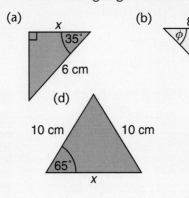

(b)

(c)

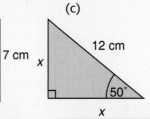

(d)

(e)

(f)

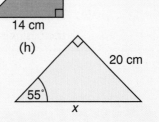

(g)

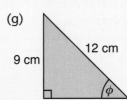

(h)

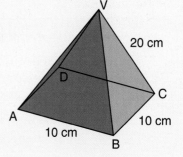

PROGRESS CHECK

2 A square-based pyramid has base ABCD and vertex V vertically above the middle of the base. AB = 10 cm and VC = 20 cm.

Find:

(a) AC
(b) the height of the pyramid
(c) the angle between VC and the base ABCD.

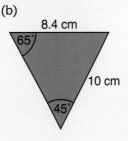

3 Find the areas of the following triangles:

(a)

(b)

(c)

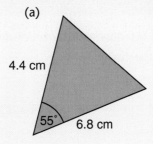

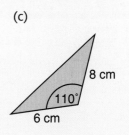

4 Find the sides and angles marked with letters in these triangles. Give angles to the nearest degree and lengths to 1 d.p.

(a)

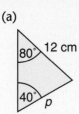

80° 12 cm

40° *p*

(b)

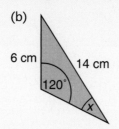

6 cm 14 cm

120°

x

(c)

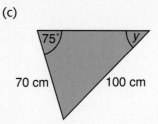

75° *y*

70 cm 100 cm

5 A yacht leaves port and sails 50 km due east. It then changes direction and sails another 50 km on a bearing of 200°.
How far from the port is the yacht now?

6 A triangle *ABC* has the following measurements:
Angle *B* = 49°; Angle *C* = 62°; side *BC* = 6.2 cm
(a) Calculate the length of side *AC*.
(b) Calculate the area of the triangle.

7 In triangle *ABC*, *AB* = 4 cm, *AC* = 6 cm and angle *A* = 53°. Calculate *BC* and angle *B*.

8 A triangle has sides of 6 cm, 7 cm and 8 cm.
Find the largest angle of the triangle.

3.5 Circles

After studying this section, you will be able to:

LEARNING SUMMARY

- **find angles in a cyclic quadrilateral**
- **find angles at the centre and circumference of a circle**
- **find angles between a radius and a chord**
- **find the angle between a tangent and a chord**
- **find the angle in the alternate segment**

Cyclic quadrilaterals

AQA A   AQA B
EDEXCEL A   EDEXCEL B
OCR A   OCR B
OCR C
NICCEA
WJEC

A **cyclic quadrilateral** is a quadrilateral drawn inside a circle so that its corners lie on the circumference of the circle.
You should know that:

(a) the **opposite** angles of a cyclic quadrilateral sum to 180°

i.e. *a* + *c* = 180°
    b + *d* = 180°

(b) the **exterior** angle of a cyclic quadrilateral is equal to the **interior opposite** angle

i.e. $e = c$

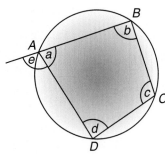

Angle theorems

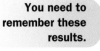

(a) The angle at the circumference subtended by a diameter is 90°. This is usually stated as 'The angle in a semicircle = 90°'.

This can be proved as follows:

The lines OA, OP and OB are equal (radii of circle).

Triangles AOP and BOP are isosceles.

Therefore in triangle APB:

$a + a + b + b = 180°$

i.e. $2(a + b) = 180°$

therefore angle $APB = a + b = 90°$

> **You need to remember these results.**

(b) The angle at the centre of a circle is twice the angle at the circumference.
Angle $AOB = 2 \times$ angle ACB

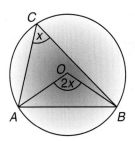

(c) Angles subtended by the same arc or chord are equal.
In Figure 1 the angles marked x are equal and subtended by chord CD, or arc CD.
In Figure 2 the angles marked x are equal, being subtended by chord BD and the angles marked y are equal, being subtended by chord AC.

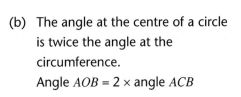

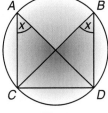

Fig 1

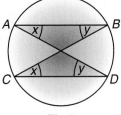

Fig 2

Examples

(a) In this diagram O is the centre of the circle. Calculate the value of angle a.

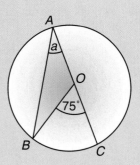

Angle $a = 37.5°$ (angle at the centre = 2 × angle at the circumference)

(b) Calculate the angles marked with letters.
angle $a = 50°$ (angles subtended by the same arc)
angle $b = 100°$ (angle at centre = twice angle at circumference)

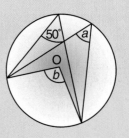

Chord theorems

NICCEA
WJEC

The line joining the mid-point of a chord to the centre of a circle is perpendicular to the chord. This can be proved using congruent triangles.

AB is a chord to the circle centre O.
X is the mid-point of AB. In triangles OXA and OXB:

$OA = OB$ (radii)
$AX = XB$ (X is the mid-point)
OX is common

∴   Triangles OXA and OXB are congruent (SSS).
∴   angle OXA = angle OXB = 90°

Tangents

NICCEA
WJEC

There are two theorems for tangents that you need to know.

(a) Tangents drawn from the same point to the same circle are equal in length. They subtend equal angles at the centre of the circle and they make equal angles with the straight line joining the centre of the circle to the point.

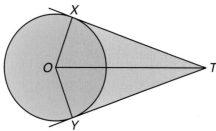

$TX = TY$
Angle XOT = angle YOT
Angle OTX = angle OTY
Angle OXT = angle OYT = 90°

(b) The angle between a tangent and a chord drawn from the point of contact is equal to any angle subtended by the chord in the alternate segment.

Angle *BXT* = angle *BAX*

This fact can be proved as follows:

Angle *ABX* = 90° (angle in a semicircle)

Angle *OXT* = 90° (angle between a diameter and a tangent)

Angle *AXB* = 90° – angle *BXT*

but angle *AXB* + angle *BAX* + angle *ABX*

= 180° (angle sum of triangle)

ie angle *AXB* + angle *BAX* = 90°

∴   90° – angle *BXT* + angle *BAX* = 90°

∴   angle *BXT* = angle *BAX*

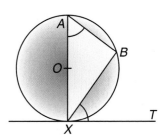

Examples

(a) Find angle *a*.

Angle *OYT* = 90°

Angle *OYX* = 40° (isosceles triangle)

∴   angle *a* = 50°

(b) Find angles *a* and *b*.

Angle *a* = 70°

Angle *b* = 50° (both are angles in alternate segment)

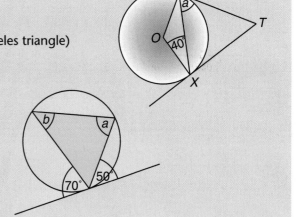

1 In each of the following diagrams *O* is the centre of the circle. Calculate the angles marked with letters.

(a)   (b)

(c)   (d)

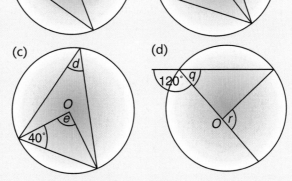

2 In each of the following diagrams O is the centre of the circle. Calculate the angles marked with letters.

(a)
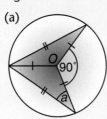

(b)
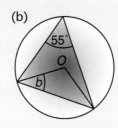

(c)
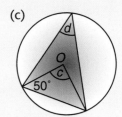

(d)
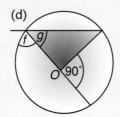

(e)
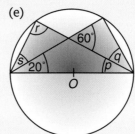

In the following questions calculate the angles marked with letters. O is the centre of each circle.

X and Y are the points of contact of the tangents to each circle.

3 (a)
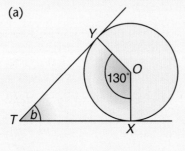

(b)
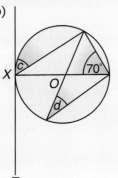

4 (a)
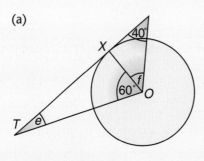

(b)
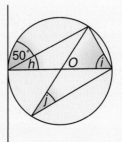

(c)
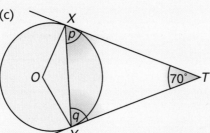

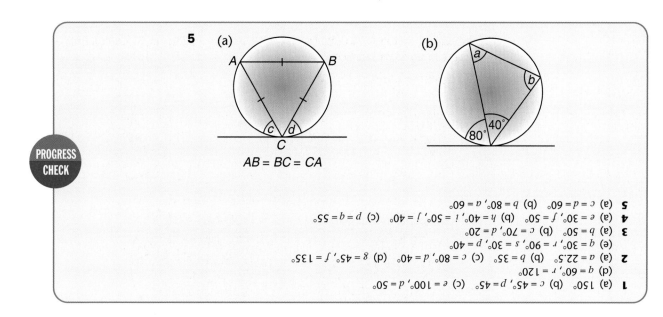

5 (a) $AB = BC = CA$

(b)

5 (a) $c = d = 60°$ (b) $b = 80°, a = 60°$
4 (a) $e = 30°, f = 50°$ (b) $h = 40°, i = 50°, j = 40°$ (c) $p = q = 55°$
3 (a) $b = 50°$ (b) $c = 70°, d = 20°$
(e) $g = 30°, r = 90°, s = 30°, p = 40°$
2 (a) $a = 22.5°$ (b) $b = 35°$ (c) $c = 80°, d = 40°$ (d) $g = 45°, f = 135°$
(d) $q = 60°, r = 120°$
1 (a) $150°$ (b) $c = 45°, p = 45°$ (c) $e = 100°, d = 50°$

3.6 3-D shapes

After studying this section, you will be able to:

- recognise a prism as a solid with a uniform cross-section
- find the volume of a prism, a pyramid, a cone and a sphere

You should know and be able to use the formulae for the areas of the following shapes: parallelogram, rhombus, trapezium. These may be needed when calculating the volumes of prisms.

Prisms and other shapes

AQA A AQA B
EDEXCEL A EDEXCEL B
OCR A OCR B
OCR C
NICCEA
WJEC

If a solid has a uniform cross-section; that is, the cross-sectional area is the same throughout its length then the solid is a prism.

You should also know the formulae for the volumes of the cone and the pyramid (although you would be given them in an examination).

Pyramid

$$\text{volume} = \frac{1}{3} \times \text{base area} \times \text{perpendicular height}$$

Note the similarity.

Cone

$$\text{volume} = \frac{1}{3} \times \text{base area} \times \text{perpendicular height}$$

$$= \frac{1}{3}\pi r^2 h$$

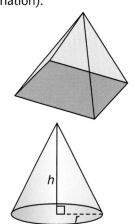

Example

A block of metal is shaped as shown below. Calculate the surface area and volume of the block.

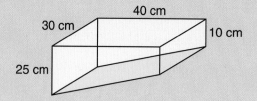

Area of trapezium at the front	$= \frac{1}{2} \times (25 + 10) \times 40$
	$= 700 \text{ cm}^2$
Area of trapezium at the back	$= 700 \text{ cm}^2$
Total area of rectangles at the side	$= 25 \times 30 + 10 \times 30$
	$= 1050 \text{ cm}^2$
Area of rectangle at the top	$= 30 \times 40$
	$= 1200 \text{ cm}^2$
Area of rectangle at base	$= 30 \times \sqrt{15^2 + 40^2}$
	$= 30 \times 42.72$
	$= 1282 \text{ cm}^2$ (to nearest whole number)
Therefore total surface area	$= 4932 \text{ cm}^2$
Volume	= cross-sectional area × length
	$= 700 \times 30$
	$= 21\ 000 \text{ cm}^3$

Example

A bucket is in the shape of a frustum of a hollow cone and is made by removing a 10 cm part of the cone as shown.
Find the volume of the bucket.

Volume of complete cone

$$= \frac{1}{3} \times \text{base area} \times \text{perpendicular height}$$

$$= \frac{1}{3} \times \pi \times 12.5^2 \times 30$$

$$= 4908.7 \text{ cm}^3$$

Volume of cone removed

$$= \frac{1}{3} \times \pi \times 5^2 \times 10$$

$$= 261.8 \text{ cm}^3$$

Volume remaining $= 4647 \text{ cm}^3$
(to the nearest whole number)

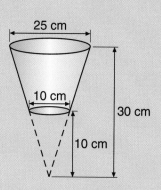

1 Find the surface area and volume of a cuboid measuring 210 mm by 100 mm by 70 mm.

2 A pyramid has a square base of side 12 cm and a height of 15 cm. Find its volume.

3 A cone has a base radius of 3.8 cm and is 8 cm high. Find its volume.

4 Find the volume of a sphere of radius 15 cm. (Volume of a sphere = $\frac{4}{3}\pi r^3$.)

1 854 cm², 1470 cm³ **2** 720 cm³ **3** 121 cm³ **4** 14 100 cm³ (3 s.f.)

3.7 Transformations

LEARNING SUMMARY

After studying this section, you will be able to:

● *specify reflections, translations, rotations and enlargements*

Reflections and translations

AQA A AQA B
EDEXCEL A EDEXCEL B
OCR A OCR B
OCR C
NICCEA
WJEC

The simplest transformations are (a) **reflections** in horizontal lines (such as $y = 0$ (the x-axis) or $y = 3$); or in vertical lines (such as $x = 0$ (the y-axis), or $x = -2$), and (b) **translations**.

Reflections

Note that reflection and translation preserve the size and shape of the object.

This diagram shows a flag, ABCDE, reflected in the lines (a) $x = 1$, (b) $y = 0$, and (c) $y = -x$

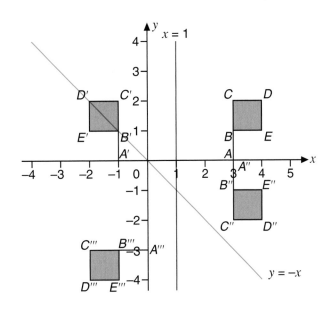

Vector notation and translations

> **KEY POINT** A vector is a quantity which has both magnitude (size), and direction.

Vectors are used to represent displacement, velocity, force, momentum, acceleration.

In geometry vectors are used to represent displacement.

In a diagram a vector can be shown in two ways:

(a) by using capital letters at each end and an arrow showing the direction.

This is the vector $\overrightarrow{AB}$.

The magnitude of the vector is written as $|\overrightarrow{AB}|$.

(b) by using a small letter and an arrow showing direction

This is the vector **a**.

In hand-writing it is shown as <u>a</u> or a̰.

Vectors can be used to represent a move from one point to another, i.e. to describe a translation.

The move represented by the vector *a* can be written as a column vector:

$$\mathbf{a} = \begin{pmatrix} x \\ y \end{pmatrix}$$

Example

The column vectors which give the translation of point A to points B, C, D, E are:

to B $\begin{pmatrix} 2 \\ 0 \end{pmatrix}$, to C $\begin{pmatrix} 0 \\ 2 \end{pmatrix}$, to D $\begin{pmatrix} -1 \\ -1 \end{pmatrix}$, to E $\begin{pmatrix} -2 \\ 3 \end{pmatrix}$.

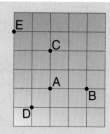

Combinations of translations can also be described using vectors.

Example

Translating the triangle ABC using the vector $\begin{pmatrix} 5 \\ 2 \end{pmatrix}$ gives the triangle $A'B'C'$. Translating triangle $A'B'C'$ using the vector $\begin{pmatrix} -2 \\ 3 \end{pmatrix}$ gives the triangle $A''B''C''$.

The translation $\begin{pmatrix} 3 \\ 5 \end{pmatrix}$ shows the movement of triangle ABC to triangle $A''B''C''$, and combining the two vectors gives $\begin{pmatrix} 5 \\ 2 \end{pmatrix} + \begin{pmatrix} -2 \\ 3 \end{pmatrix} = \begin{pmatrix} 3 \\ 5 \end{pmatrix}$.

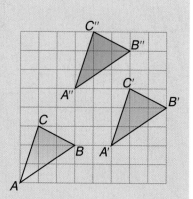

Rotations

AQA A   AQA B
EDEXCEL A   EDEXCEL B
OCR A   OCR B
OCR C
NICCEA
WJEC

> By convention an anti-clockwise rotation is positive and a clockwise rotation is negative.

> Rotation preserves shape and size so the rotated shapes are congruent to the original shape.

To describe a rotation requires three pieces of information:

(a) the angle of the rotation;   (b) the direction of the rotation;   (c) the centre of the rotation.

Example

(a) rotating triangle *ABC* through −90°, i.e. clockwise, about (0, 0) gives the image *A'B'C'*.

(b) rotating triangle *A'B'C'* through 180° about (0, 0) gives *A"B"C"*.

(c) It is possible to define a single transformation to rotate *ABC* to *A"B"C"*: a rotation of 90° about (0, 0).

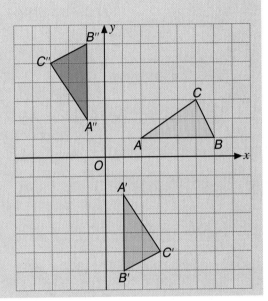

It is a straightforward process to rotate a shape through any size of angle:

Example

Rotate triangle *ABC* through 40° about the point *O*. Label the image *A'B'C'*.

Draw lines *OA*, *OB*, *OC*. Measure the distances *OA*, *OB*, *OC*; draw lines *OA'*, *OB'*, *OC'* so that *OA* = *OA'*, *OB* = *OB'*, *OC* = *OC'*, and the angles *AOA'* = *BOB'* = *COC'* = 40°. Alternatively measure *OA* and draw *OA'* = *OA* and *AOA'* = 40° and use tracing paper to complete the image.

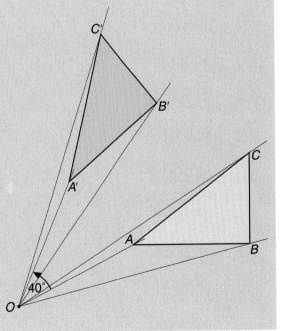

To find the centre of rotation you need to know the position of a point on the object, say A, and its image, A'. The centre of rotation must be the same distance from A and A'

Join A to A' and construct the perpendicular bisector of the line AA'. The centre of rotation, O, will lie on this line, but, to fix the centre requires the process to be repeated for another point and its image, i.e. B and B'.

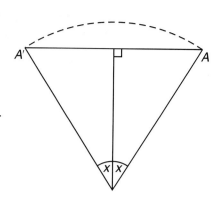

Thus for an object, such as triangle ABC, and its image, triangle $A'B'C'$, the centre of rotation is found as follows:

(a) join A to A' and B to B'

(b) draw the perpendicular bisectors of lines AA' and BB'

(c) the point where these bisectors intersect is the centre of rotation, O.

This method can be used to find the single equivalent rotation which replaces two or more rotations.

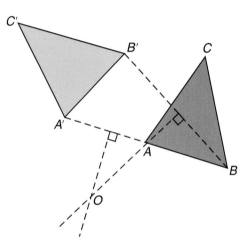

Example

Rotate triangle ABC through $30°$ about the origin. Label its image $A'B'C'$.
Rotate triangle $A'B'C'$ through $70°$ about the point P, $(0, 2)$. Label this image $A''B''C''$.

What single rotation will map triangle ABC onto triangle $A''B''C''$?

> **Draw the perpendicular bisector of AA″ and CC″ They intersect at the centre of rotation (0.4, 1.6).**

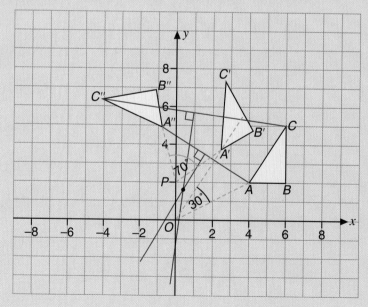

The single rotation is $103°$ about the point (0.4, 1.6).

Enlargements

To carry out an enlargement requires two pieces of information:

(a) the scale factor (b) the centre of the enlargement.

Example

Rectangle *ABCD* is enlarged by a scale factor of 2; the centre of enlargement is the origin.

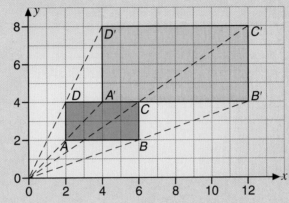

This means that $OA' = 2 \times OA$ measured along OA extended

$OB' = 2 \times OB$ measured along OB extended

$OC' = 2 \times OC$ measured along OC extended

$OD' = 2 \times OD$ measured along OD extended

The image *A'B'C'D'* is four times the size of the object *ABCD*.

If the scale factor is negative the image is on the opposite side of the centre of enlargement, and the image is inverted, as shown in the next example.

Example

Triangle *A*(2, 2), *B*(4, 2), *C*(2, 5) is enlarged by a scale factor −2. The centre of enlargement is the point *P*, (0, 1). This is shown in the diagram where, for example $PA' = 2 \times PA$ as measured along the line *AP* extended, that is to the left of *P*.

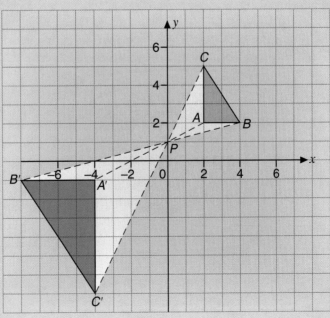

PROGRESS CHECK

1 Draw a triangle ABC at $A(1,2)$, $B(3,5)$, $C(6,2)$.
  (a) Find the image of triangle ABC under a 90° rotation, anticlockwise about the origin.
  (b) Find the image of triangle ABC under a 90° rotation, clockwise about the point $(2,-2)$.

2 Draw a triangle with vertices $A(1,2)$, $B(1,6)$ $C(3,5)$. Rotate this triangle through 70° about the point $(1,2)$.

3 Draw a triangle $A(4,4)$, $B(1,4)$, $C(1,2)$ and rotate it through 40° about the origin giving triangle $A'B'C'$. Rotate triangle $A'B'C'$ through $-70°$ about the point $(-1,-1)$.
  What single rotation is equivalent to the combination of these two rotations?

4 Draw a set of axes, the x-axis from -16 to $+6$, the y-axis from -6 to $+4$. Draw a quadrilateral with coordinates $A(2,2)$, $B(5,0)$, $C(5,-1)$, $D(2,-1)$. Draw the enlargement of $ABCD$ by a scale factor of -3 with centre of enlargement, the origin.

1 (a) $A'(-2,1)$, $B'(-5,3)$, $C'(-2,6)$
  (b) $A'(6,-1)$, $B'(9,-3)$, $C'(6,-6)$
2 $A'(1,2)$, $B'(-2.8,3.3)$, $C'(-1.1,4.8)$
3 31° about the point $(-1,1,3)$
4 $A'(-6,-6)$, $B'(-15,0)$, $C'(-15,3)$, $D'(-6,3)$

3.8 Area and volume

LEARNING SUMMARY

After studying this section, you will be able to:

- **distinguish between formulae for length, area and volume**
- **find areas and volumes of similar figures**

Dimensions

AQA A   AQA B
EDEXCEL A   EDEXCEL B
OCR A   OCR B
OCR C
NICCEA
WJEC

You need to be able to distinguish between the formulae for length, area and volume.

A formula involving one dimension, such as height, is a measurement of length, e.g. cm.

A formula involving two dimensions, such as width × length, is a measurement of area, e.g. cm^2.

A formula involving three dimensions, such as length × width × height, is a measurement of volume, e.g. cm^3.

> **Example**
> What quantities do the following represent?
>
> (a) $\dfrac{4}{3}\pi r^3$ (b) $2\pi rh$
>
> (a) $\dfrac{4}{3}\pi r^3$ includes $r \times r \times r$ i.e. three lengths multiplied together.
>
> Therefore it represents a volume.
>
> (b) $2\pi rh$ includes $r \times h$ i.e. a length × length, therefore it represents an area.

Areas and volumes of similar figures

AQA A AQA B
EDEXCEL A EDEXCEL B
OCR A OCR B
OCR C
NICCEA
WJEC

If two figures are similar then the corresponding edges on those figures are in the same ratio.

A 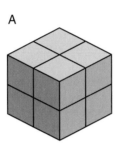 B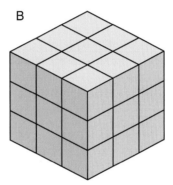

The ratio of the lengths of the sides of these two cubes is 2 : 3.

The area of a face of cube A is 4 square units.
The area of a face on cube B is 9 square units
The areas of the faces are in the ratio 4 : 9 ie $2^2 : 3^2$.
The total surface area of cube A is 6 × 4 square units = 24 square units, that for cube B is 6 × 9 square units ie 54 square units.
The ratio of the surface areas is 24 : 54 = 4 : 9 = $2^2 : 3^2$.
Thus it can be seen that the ratio of the areas of these similar figures is the ratio of the squares of the lengths of the sides.
In general terms:

if the ratio of the lengths is $a : b$ then the ratio of the areas is $a^2 : b^2$

The volume of cube A = 2 × 2 × 2 = 8 cubic units
The volume of cube B = 3 × 3 × 3 = 27 cubic units
The ratio of the volumes = 8 : 27 = $2^3 : 3^3$.
In the same way, if the ratio of the lengths is $a : b$, then the ratio of the volumes is $a^3 : b^3$.

Example

Two similar cones are made, the larger cone has a diameter twice that of the smaller cone. The smaller cone has a surface area of 500 cm² and a volume of 130 cm³. Calculate the surface area and volume of the larger cone.

Ratio of lengths = 2 : 1
∴   Ratio of areas = $2^2 : 1^2 = 4 : 1$
Thus surface area of larger cone = 4 × 500 = 2000 cm².
Ratio of volumes = $2^3 : 1^3 = 8 : 1$
Thus volume of larger cone = 8 × 130 cm³ = 1040 cm³

PROGRESS CHECK

1 Which of the following represent areas and which volumes?

(a) $\frac{1}{2}bh$ (b) $\frac{1}{2}ab\sin C$ (c) $2\pi r^2(r+h)$ (d) $\frac{4}{3}\pi(r^2+h^2)$ (e) $2\pi r^2 + 2\pi rh$

2 Two similar cylinders have heights of 3 cm and 6 cm respectively. If the volume of the smaller cylinder is 30 cm³ find the volume of the larger cylinder.

3 Two spheres made of the same metal have weights of 32 kg and 108 kg. If the radius of the larger sphere is 9 cm find the radius of the small sphere (assume weights are proportional to volumes).

1 (a) area (b) area (c) volume (d) area (e) area
2 240 cm³ **3** 6 cm

3.9 **Vectors**

After studying this section, you will be able to:

LEARNING SUMMARY

● **identify the zero vector and unit vectors**
● **understand equal and inverse vectors**
● **know what a scalar is**
● **add and subtract vectors**
● **solve problems using vectors**

Zero vector and unit vectors

AQA A   AQA B
EDEXCEL A   EDEXCEL B
OCR A   OCR B
OCR C
NICCEA
WJEC

Vectors were introduced earlier when looking at translations.

A vector with magnitude 0 is called the **zero vector**, written **0**. A vector with magnitude 1 is called a **unit vector**.

Vectors are equal if they have the same magnitude and the same direction.

a = b

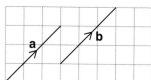

Inverse vectors

The **inverse** of a vector is a vector of equal magnitude but in the opposite direction. The inverse of $\overrightarrow{AB}$ is $-\overrightarrow{AB}$ or $\overrightarrow{BA}$ and the inverse of **a** is $-$**a**.

Scalars

Scalars are quantities which have magnitude but not direction. You can multiply a vector by a scalar to produce another vector.

For **example**, multiplying vector **a** by **2** gives a vector twice the length of **a** and parallel to **a**.

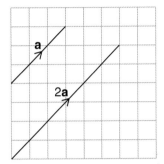

If $\mathbf{a} = \begin{pmatrix} 3 \\ 3 \end{pmatrix}$ then $2\mathbf{a} = 2\begin{pmatrix} 3 \\ 3 \end{pmatrix} = \begin{pmatrix} 6 \\ 6 \end{pmatrix}$

Vector addition and subtraction

When two vectors are added or subtracted to produce a third vector this vector is called the **resultant**. The resultant vector is marked with a double arrowhead.

Triangle law

To add two vectors means apply the first vector then apply the second vector.
$$\overrightarrow{AB} + \overrightarrow{BC} = \overrightarrow{AC}$$
or **a** + **b** = **c**
This is known as the **triangle law**.

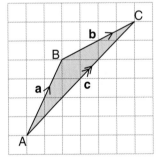

Parallelogram law

The **parallelogram law** shows that going from A to C via B is the same as going from A to C via D.
In other words:
$\overrightarrow{AB} + \overrightarrow{BC} = \overrightarrow{AC}$ is the same as
$\overrightarrow{AD} + \overrightarrow{DC} = \overrightarrow{AC}$
or **a** + **b** = **b** + **a** = **c**

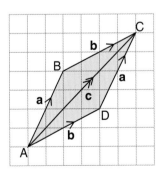

Subtracting a vector is the same as adding its inverse:
a − **b** is the same as **a** + (−**b**).

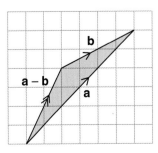

Vectors can be used to prove geometric properties and relationships

> **Example**
> In the triangle ABC the points X and Y are the mid-points of AB and AC.
> Show that line XY is parallel to BC and half its length.
>
> Let $\vec{AX} = \mathbf{a}$ and $\vec{AY} = \mathbf{b}$
> $$\therefore \quad \vec{XY} = \vec{XA} + \vec{AY}$$
> $$= -\mathbf{a} + \mathbf{b} = \mathbf{b} - \mathbf{a}$$
> $$\vec{AX} = \tfrac{1}{2}\vec{AB}$$
> so $\quad \vec{AB} = 2\vec{AX}$
> $$= 2\mathbf{a}$$
> $$\vec{AY} = \tfrac{1}{2}\vec{AC}$$
> so $\quad \vec{AC} = 2\vec{AY} = 2\mathbf{b}$
> $$\vec{BC} = \vec{BA} + \vec{AC}$$
> $$= -2\mathbf{a} + 2\mathbf{b}$$
> $$= 2(\mathbf{b} - \mathbf{a})$$
> $$\therefore \quad \vec{BC} = 2\vec{XY} \text{ or } \vec{XY} = \tfrac{1}{2}\vec{BC}$$
> $\therefore$ XY is parallel to BC and half its length.

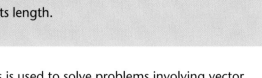

Finding the resultant of two vectors is used to solve problems involving vector quantities such as force and velocity. Note that as force is a vector quantity then both magnitude and direction should be given.

> **Examples**
> (a) Find the resultant of two forces of magnitude 5 N and 12 N acting on a mass as shown in the diagram.
>
>
>
> Add the two forces as vectors, joining them 'nose to tail'.
>
>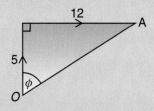
>
> The magnitude of the resultant is found using Pythagoras and is 13 N.
> Find the angle using trigonometry.
> $$\tan \phi = \frac{12}{5}$$
> $$\therefore \quad \phi = 67.4°$$
> The resultant is a force of 13 N at 67.4° to the direction of the 5 N force.

(b) An aircraft can fly at 300 mph in still air. The wind is blowing at 40 mph towards the south-east. If the aircraft heads due north what is the actual velocity relative to the ground?

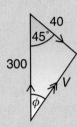

Actual velocity = velocity of aircraft + velocity of the air
Using cosine rule:

$$V^2 = 300^2 + 40^2 - 2 \times 300 \times 40 \times \cos 45°$$
$$= 273 \text{ mph}$$

Using sine rule $\dfrac{\sin \phi}{40} = \dfrac{\sin 45}{273}$

∴ $\phi = 6°$

Velocity is 273 mph at 006°.

PROGRESS CHECK

1 OABC is a quadrilateral, $\overrightarrow{OA} = \mathbf{a}$, $\overrightarrow{OB} = \mathbf{b}$ and $\overrightarrow{OC} = \mathbf{c}$.

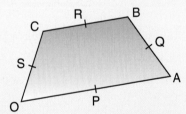

P, Q, R, S are the mid-points of OA, AB, BC and OC respectively.

(a) Find, in terms of **a**, **b** and **c**:
　　(i) $\overrightarrow{OP}$　(ii) $\overrightarrow{AB}$　(iii) $\overrightarrow{AQ}$　(iv) $\overrightarrow{PQ}$　(v) $\overrightarrow{SR}$
(b) Prove that PQ is parallel to SR.
(c) What type of quadrilateral is PQRS?

2 Two forces, P and Q, act on a mass. The resultant of P and Q is a force of magnitude 10 N and acts on a bearing of 055°. If force P has a magnitude of 12 N and acts due north find the magnitude and bearing of force Q.

2 10.3 N, 127.4°

(c) a parallelogram

1 (a) (i) $\frac{1}{2}$**a** (ii) (**b** − **a**) (iii) $\frac{1}{2}$(**b** − **a**) (iv) $\frac{1}{2}$**b** (v) $\frac{1}{2}$**b**

3.10 Measures and constructions

After studying this section, you will be able to:

- **understand that measurements are approximate**
- **work with compound units**
- **do simple constructions**

Upper and lower bounds

AQA A   AQA B
EDEXCEL A   EDEXCEL B
OCR A   OCR B
OCR C
NICCEA
WJEC

All measurements are approximations. Measurements are given to the nearest practical unit. Measuring a value to the nearest unit means deciding that it is nearer to one mark on a scale than another; in other words it is within half a unit of that mark.

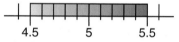

4.5   5   5.5

Anything within the shaded area is 5 to the nearest unit. The boundaries for this interval are 4.5 and 5.5 and this would be written as 4.5 ≤ measurement < 5.5. where 4.5 is the lower bound and 5.5 is the upper bound.

Examples

(a) Tom won the 100 m race with a time of 12.2 seconds, to the nearest tenth of a second. What are the upper and lower bounds for this time?

The lower bound = 12.15 sec and the upper bound = 12.25 sec.

(b) A mass, given as 46 kg to the nearest kg, lies between what limits?
The mass is between 45.5 kg and 46.5 kg.

Compound units

AQA A   AQA B
EDEXCEL A   EDEXCEL B
OCR A   OCR B
OCR C
NICCEA
WJEC

Some measures depend upon other measures, for example:

$$\text{average speed} = \frac{\text{total distance travelled}}{\text{total time taken}}, \text{ and density} = \frac{\text{mass}}{\text{volume}}.$$

Examples

(a) Find the average speed of a car which travelled 150 miles in two and a half hours.

$$\text{average speed} = \frac{150}{2.5} = 60 \text{ mph}$$

(b) Calculate the density of a rock of mass 780 g and volume 84 cm^3. Give the answer to a suitable degree of accuracy.

$$\text{density} = \frac{780}{84} = 9.28571 \text{ g cm}^{-3} = 9 \text{ g cm}^{-3}$$

Constructions

AQA A   AQA B
EDEXCEL A   EDEXCEL B
OCR A   OCR B
OCR C
NICCEA
WJEC

You need to be able to complete various constructions. Remember it is often helpful to make a sketch diagram first.

Triangles

The standard constructions for triangles were outlined on pages 86 and 87 where the properties of congruent triangles were discussed.

The perpendicular bisector of a line

To bisect line AB:

- using a pair of compasses, with centres A and B, draw arcs with the same radius to intersect either side of line AB
- join the points of intersection: this line is the perpendicular bisector of AB.

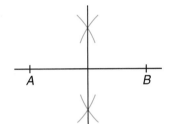

The perpendicular from a point to a line

To draw the perpendicular from P to a given line:

- from P draw arcs to cut the line at A and B
- from A and B draw arcs with the same radius to intersect at C
- Join P to C: this line is perpendicular to the line AB.

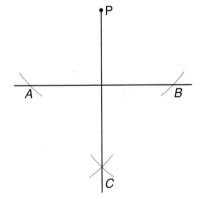

To bisect an angle

To bisect the angle at A:

- with centre A draw arcs to cut the lines at B and C
- with the same radius draw arcs centre B and C to cut at D
- join A to D.

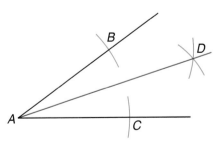

1   Complete the sentences:
    (a) A height given as 5.7 m to two significant figures is between ___ m and ___ m.
    (b) A volume given as 568 ml, to the nearest millilitre, is between ___ ml and ___ ml.
    (c) A winning time given as 23.93 s to the nearest hundredth of a second is between ___ s and ___ s.

2   A car travels 20 km in 12 minutes. What is the average speed in km/h?

3   Calculate the density of a stone of mass 350 g and volume 45 cm^3.

1 (a) 5.65 m and 5.75 m   (b) 567.5 ml and 568.5 ml
(c) 23.925 s and 23.935 s
2 100 km/h
3 7.7 g cm^{-3}

3.11 Mensuration

After studying this section, you will be able to:

● **find the volume and surface area of a sphere**
● **find the length of an arc**
● **find the area of a sector and a segment**

The sphere

AQA A   AQA B
EDEXCEL A   EDEXCEL B
OCR A   OCR B
OCR C
NICCEA
WJEC

KEY
POINT

The volume of a sphere $= \dfrac{4}{3}\pi r^3$

The surface area of a sphere $= 4\pi r^2$

You should have completed the work on volumes of 3D shapes in Section 3.6.
Here are some further examples and questions to attempt.

Examples

(a) Find the volume and surface area of a sphere with a radius of 10 cm.

$$\text{Volume} = \frac{4}{3}\pi r^3$$

$$= \frac{4}{3} \times 3.142 \times 10^3$$

$$= 4189 \text{ cm}^3 \text{ (to the nearest whole number)}$$

$$\text{Surface area} = 4\pi r^2$$

$$= 4 \times 3.142 \times 10^2$$

$$= 1257 \text{ cm}^2 \text{ (to the nearest whole number)}$$

(b) A solid metal cone of height 20 cm and radius 12 cm is melted down to form a cylinder of the same height. What is the radius of the cylinder?

$$\text{Volume of cone} = \frac{1}{3}\pi r^2 h$$

$$= \frac{1}{3} \times \pi \times 12^2 \times 20$$

$$\text{Volume of cylinder} = \pi r^2 h$$

$$= \pi \times r^2 \times 20$$

$$\therefore \quad \frac{1}{3} \times \pi \times 12^2 \times 20 = \pi \times r^2 \times 20$$

$$\therefore \quad r^2 = \frac{1}{3} \times 12^2$$

$$\text{so} \quad r^2 = 48$$

$$r = 6.9 \text{ cm}$$

Arcs, sectors and segments

AQA A   AQA B
EDEXCEL A   EDEXCEL B
OCR A   OCR B
OCR C
NICCEA
WJEC

Arcs

An **arc** is part of the circumference of a circle. The ends of an arc are formed by two radii as shown in the diagram.

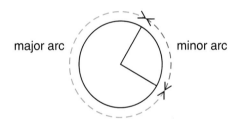

major arc   minor arc

The arc which subtends the larger angle at the centre is called the **major arc**, the arc which subtends the smaller angle at the centre is the **minor arc**.

The length of the arc depends on the size of the angle at the centre and on the radius of the circle. If the angle turned through to produce the arc is θ, then arc length $= \frac{\theta}{360} \times \pi d$ or $\frac{\theta}{360} \times 2\pi r$.

Example

Calculate the length of the minor arc of a circle radius 6 cm if the angle formed at the centre is 75°.

$$\text{Length of arc} = \frac{\theta}{360} \times 2\pi r$$

$$= \frac{75}{360} \times 2 \times 3.142 \times 6$$

$$= 7.9 \text{ cm to 2 s.f.}$$

Sectors

A **sector** of a circle is an area bounded by two radii and an arc. A **minor sector** is formed by a minor arc, a **major sector** by a major arc.

In the same way as the arc length is a fraction of the circumference of the circle so the area of a sector is a fraction of the area of the circle.

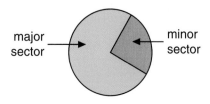

major sector · minor sector

> **KEY POINT**
>
> **Area of sector** $= \dfrac{\theta}{360} \times \pi r^2$

Example

Calculate the area of the sector of a circle with subtended angle 70° and radius 10 cm.

$$\text{Area of sector} = \frac{\theta}{360} \times \pi r^2$$

$$= \frac{70}{360} \times 3.142 \times 10^2$$

$$= 61.1 \text{ cm}^2$$

Segments

A **segment** of a circle is the area bounded by a chord and an arc.

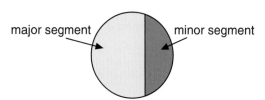

major segment · minor segment

To work out the area of a minor segment first work out the area of the sector, $AOBC$, and then subtract the area of the triangle AOB formed by the chord and the two radii.

For the major segment, the area is found by calculating the area of the major sector, $AOBD$, and adding the area of the triangle AOB.

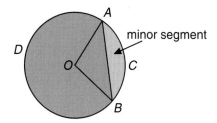

minor segment

Example

To help to prevent liquid spillages from spreading a barrier is made from a rubber compound.

The cross-section of the barrier is as shown below. It is made from a circular tube, of radius 20 cm, with a flat base. Calculate the volume of rubber needed for a 10 m length of tubing.

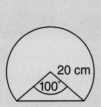

$$\text{Area of major sector} = \frac{260}{360} \times \pi \times 20^2$$

$$= 907.6 \text{ cm}^2$$

Using the formula
$$A = \frac{1}{2} ab \sin C$$

$$\text{Area of triangle} = \frac{1}{2} \times 20 \times 20 \times \sin 100$$

$$= 197 \text{ cm}^2$$

$$\text{Total cross-sectional area} = 1104.6 \text{ cm}^2$$

$$\therefore \quad \text{Volume} = 10 \times 100 \times 1104.6 \text{ cm}^3$$

$1 \text{ m}^3 = 10^6 \text{ cm}^3$

$$= 1\,104\,600 \text{ cm}^3 = 1.1 \text{ m}^3$$

PROGRESS CHECK

1 A child's toy is made from a cylinder, of radius 3 cm and height 3 cm with a hemisphere fixed to the top, as shown.
   Find the volume of the toy.

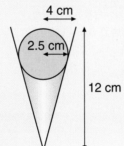

2 The sketch shows an ice-cream cone of radius 4 cm and depth 12 cm in which there is a sphere of ice cream of radius 2.5 cm. The ice cream melts and runs into the base of the cone. Find the depth of the liquid ice cream when this has happened.

3 Find the shaded area in the following diagram:

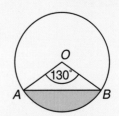

4 The chord AB subtends an angle of 130° at the centre, O, of a circle.
   The diameter of the circle is 16 cm.
   Find the area of the minor segment, which is shaded in the diagram.

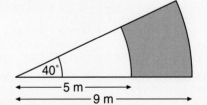

1 141 cm³ to the nearest whole number
2 8.25 cm
3 19.5 m²
4 48.1 cm²

3.12 Loci

LEARNING SUMMARY

After studying this section, you will be able to:

● construct a locus which fulfils specified conditions

Three important loci

AQA A   AQA B
EDEXCEL A   EDEXCEL B
OCR A   OCR B
OCR C
NICCEA
WJEC

KEY POINT   The word locus describes the position of points which obey a certain rule.

Three important loci are:

(a) The circle: the locus of points which are equidistant from a fixed point, the centre.

(b) The perpendicular bisector: the locus of points which are equidistant from two fixed points A and B (see section 3.10).

(c) The angle bisector: the locus of points which are equidistant from two fixed lines (see section 3.10).

Example

The diagram shows the walls of a rectangular shed, ABCD, measuring 8 m by 5 m. A goat is tied to the corner C of the shed by a rope 6 m long. Show the part of the shed that the goat can reach.

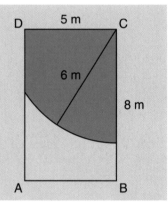

PROGRESS CHECK

1 Draw a triangle ABC where AB = 9 cm and BC = 7 cm, and AC = 5 cm.
   (a) Show the locus of points within the triangle which are equidistant from AC and AB.
   (b) Show the locus of points within the triangle which are less than 5 cm from B and nearer to AC than to AB.

2 Two radio stations, A and B, 80 km apart, broadcast over distances of 55 km and 65 km respectively.
   Using a scale of 1 cm to 20 km show the area where both stations can be heard.

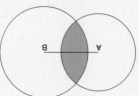

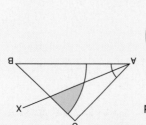

2 The required area is shaded.
1 (a) The locus is the bisector of angle CAB shown as line AX.
   (b) the area is inside the arc drawn centre B, radius 5 cm and above the line AX – shown shaded in the diagram.

Sample GCSE questions

1 This coil of piping has 15 turns.
The diameter of the coil is about 3 m.

(a) Estimate the length of the pipe. **[2]**
(b) This is the cross-section of the pipe.

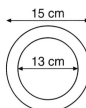

15 cm

13 cm

A one-metre length has mass 0.89 kg.
Calculate the density of the material.
State the units of your answer. **[7]**

This is an example of a multistep question. You are not asked specifically to find the volume but this is a necessary step towards finding the density.

(a) $\pi \times 3 \times 15$ ✔
$\approx 140\ m$ (or 135 m) ✔

(b) Area of cross-section $= \pi(7.5^2 - 6.5^2)\,cm^2$ ✔✔
Volume of 1-metre length $= 100\pi(7.5^2 - 6.5^2)\,cm^3$ ✔
$= \dfrac{\pi(7.5^2 - 6.5^2)}{10\,000}\,m^3$ ✔
$= 0.0044\ m^3$

Divide by 1 000 000 to change cm³ to m³.

$Density = \dfrac{mass}{volume}$ ✔

You have to state the units. You could give the answer in g/cm³. This is the given answer divided by 1000.

$= \dfrac{0.89}{0.0044} = 202.3\ kg/m^3$ ✔✔

2 $ABCD$ is an isosceles trapezium. Angle $ACD = x°$.

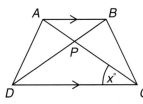

(a) Find the size of:
(i)   angle ABD
(ii) angle APD
giving your reasons. **[3]**

(b) Prove that $ABCD$ is a cyclic quadrilateral **[4]**

Sample GCSE questions

This is a straightforward question designed to test your ability to reason geometrically and your knowledge of angle facts etc.

It may help to mark on the diagram the facts you are given and the values that you calculate as you work towards the solution.

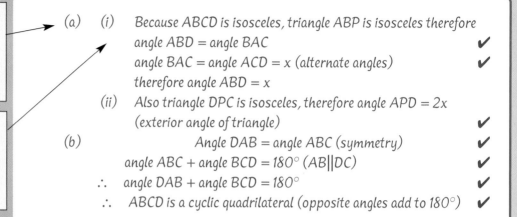

(a) (i) Because ABCD is isosceles, triangle ABP is isosceles therefore
angle ABD = angle BAC ✔
angle BAC = angle ACD = x (alternate angles) ✔
therefore angle ABD = x

(ii) Also triangle DPC is isosceles, therefore angle APD = 2x
(exterior angle of triangle) ✔

(b) Angle DAB = angle ABC (symmetry) ✔
angle ABC + angle BCD = 180° (AB∥DC) ✔
∴ angle DAB + angle BCD = 180° ✔
∴ ABCD is a cyclic quadrilateral (opposite angles add to 180°) ✔

3 The diagram shows a closed tank used to mix ice-cream.

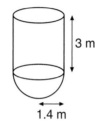

3 m

1.4 m

The tank is in the form of a cylinder of height 3 m fixed on top of a hemisphere of radius 1.4 m.

(a) Calculate, to the nearest square metre, the total surface area of the tank. **[4]**

The tank contains ice cream to a level 30 cm below the top.

(b) Calculate, in cubic metres, correct to one decimal place, the volume of ice cream in the tank. **[4]**

The ice cream is poured into tubs. The tubs are in the form of a cone of height $(h + 7)$cm, diameter 8 cm with a cone of height h cm and diameter 6 cm removed.

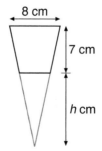

8 cm

7 cm

h cm

(c) (i) Using similar triangles, calculate the height h. **[3]**

(ii) Calculate the number of tubs which can be filled with ice-cream from the tank using your answer from part (b). **[4]**

Sample GCSE questions

Remember to divide the area and volume formula for a sphere by 2 in this question.

Your answers may differ slightly depending on the value of π used.

Although your answers are rounded use unrounded values in subsequent calculations.

(a) Surface area $= \pi r^2 + 2\pi rh + 2\pi r^2$ ✔
$$= \pi \times 1.4^2 + 2 \times \pi \times 1.4 \times 3 + 2 \times \pi \times 1.4^2$$
$$= 6.158 + 26.393 + 12.317$$ ✔
$$= 44.87 \ m^2$$ ✔
$$= 45 \ m^2$$ ✔

(b) $3 - 0.3 = 2.7$ ✔
$$Volume = \pi \times 1.4^2 \times 2.7 + \frac{1}{2} \times \frac{4\pi}{3} \times 1.4^3$$ ✔
$$= 16.63 + 5.75$$ ✔
$$= 22.38 = 22.4 \ m^3 \ to \ 1 \ d.p.$$ ✔

(c) (i) $\dfrac{h}{h+7} = \dfrac{3}{4}$ ✔
$$\therefore \quad 4h = 3h + 21$$ ✔
$$\therefore \quad h = 21$$ ✔

(ii) $Volume \ of \ tub = \dfrac{1}{3} \times \pi \times 4^2 \times (21+7) - \dfrac{1}{3} \times \pi \times 3^2 \times 21$ ✔
$$= \frac{1}{3} \times \pi (16 \times 28 - 9 \times 21)$$ ✔
$$= 271.26 \ cm^3$$ ✔
$$\therefore \quad number \ of \ tubs = \frac{22.38 \times 1\,000\,000}{271.26}$$
$$= 82503.9 = 82\,500 \ to \ the \ nearest \ hundred$$ ✔

4   The shaded segment of the diagram shows the shape of the interior window ledge of a large bay window for a shop.

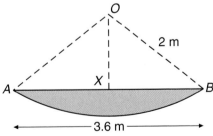

The straight edge AB is 3.6 m long and the curved edge AB is an arc of a circle of radius 2 m.

The window ledge is to be covered with a sheet of plastic with a length of beading curved along the arc AB.

(a) Calculate:
(i)   the angle AOB [3]
(ii)   the length of beading.   [3]

(b) Calculate the area of the window ledge.   [8]

Sample GCSE questions

(a) (i) $\sin XOB = \dfrac{1.8}{2}$ ✔

angle $XOB = 64.16°$ ✔
angle $AOB = 128.3°$ ✔

(ii)   length of arc $= \dfrac{128.3}{360} \times 2 \times \pi \times 2$ ✔✔

$= 4.48\ m$ ✔

(b)   area of sector $= \dfrac{128.3}{360} \times \pi \times 2^2$ ✔

$= 4.48\ m^2$ ✔
$OX = \sqrt{2^2 - 1.8^2}$ ✔
$= 0.87\ m$ ✔

You could have used
$A = \dfrac{1}{2}\ ab \sin C$ *to find the area of triangle OAB.*

i.e.

$A = \dfrac{1}{2} \times 2 \times 2 \times \sin 128.4°$

area of triangle $OAB = \dfrac{1}{2} \times 3.6 \times 0.87$ ✔

$= 1.57\ m^2$ ✔
area of ledge $= 4.48 - 1.57$
$= 2.91\ m^2$ ✔
$= 2.9\ m^2$ (to 1 d.p.)   ✔

5   ABCD is a parallelogram. M is the mid-point of DC.

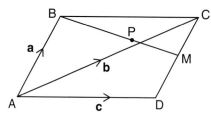

(a)   Find $\overrightarrow{DC}$, $\overrightarrow{DM}$, $\overrightarrow{AM}$ in terms of **b** and **c**.   **[3]**

(b)   Find $\overrightarrow{BM}$ in terms of **a**, **b** and **c**.   **[2]**

P is a point on BM such that $\overrightarrow{BP} = \dfrac{2}{3}\ \overrightarrow{BM}$.

(c)   Express $\overrightarrow{BP}$ in terms of **a** and show that $\overrightarrow{AP}$ can be expressed
in the form $\dfrac{1}{n}\,(\mathbf{a} + \mathbf{b} + \mathbf{c})$ where n is an integer.   **[5]**

(d)   Use the fact that **b** = **a** + **c** to find $\overrightarrow{AP}$ in terms of **b** only.
What can you deduce about the position of P?   **[3]**

Sample GCSE questions

Make sure you check the directions of the vectors you are using.

(a) $\vec{DC} = \vec{DA} + \vec{AC} = -\mathbf{c} + \mathbf{b}$ ✔

$\vec{DM} = \dfrac{1}{2}\vec{DC} = \dfrac{1}{2}(-\mathbf{c} + \mathbf{b})$ ✔

$\vec{AM} = \vec{AD} + \vec{DM} = \mathbf{c} + \dfrac{1}{2}(-\mathbf{c} + \mathbf{b}) = \dfrac{1}{2}(\mathbf{c} + \mathbf{b})$ ✔

(b) $\vec{BM} = \vec{BA} + \vec{AM}$

There are alternative ways of reaching the solution i.e. in part (b) $\vec{BM}$ also equals $\vec{BC} + \vec{CM}$.

$= -\mathbf{a} + \dfrac{1}{2}(\mathbf{c} + \mathbf{b})$ ✔

$= -\mathbf{a} + \dfrac{1}{2}\mathbf{c} + \dfrac{1}{2}\mathbf{b}$ ✔

(c) $\vec{BP} = \vec{BA} + \vec{AP}$

$= -\mathbf{a} + \vec{AP}$

$\therefore \vec{AP} = \vec{BP} + \mathbf{a}$ ✔

But $\vec{BP} = \dfrac{2}{3}\vec{BM}$

$= \dfrac{2}{3}\left(-\mathbf{a} + \dfrac{1}{2}\mathbf{c} + \dfrac{1}{2}\mathbf{b}\right)$ ✔

$\therefore \vec{AP} = -\dfrac{2}{3}\mathbf{a} + \dfrac{1}{3}\mathbf{b} + \dfrac{1}{3}\mathbf{c} + \mathbf{a}$ ✔

$= \dfrac{1}{3}\mathbf{a} + \dfrac{1}{3}\mathbf{b} + \dfrac{1}{3}\mathbf{c} = \dfrac{1}{3}(\mathbf{a} + \mathbf{b} + \mathbf{c})$ ✔

$\therefore n = 3$ ✔

(d) $\vec{AP} = \dfrac{1}{3}(\mathbf{a} + \mathbf{b} + \mathbf{c}) = \dfrac{1}{3}[(\mathbf{a} + \mathbf{c}) + \mathbf{b}]$

$= \dfrac{1}{3}(\mathbf{b} + \mathbf{b})$ ✔

$= \dfrac{2}{3}\mathbf{b}$ ✔

Therefore P is on the line AC and $\dfrac{2}{3}$ of the way along it. ✔

Exam practice questions

1

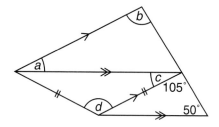

(a) Calculate the following angles:
   (i) b **[1]**
   (ii) c **[1]**
   (iii) a **[1]**
   (iv) d **[2]**

(b) In the quadrilateral PQRS, X is the mid-point of QR, PX is parallel to SR and SX is parallel to PQ.

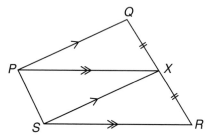

Prove that triangle PQX is congruent to triangle SXR.   **[2]**

2 A hemispherical bowl, radius r, is partially filled with water to a depth of $\dfrac{r}{2}$, as shown in the diagram.

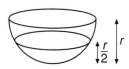

Which of the following formulae could give the volume of water in the bowl?
Give a reason for your choice.

(a) $\dfrac{\pi r^2}{4}$ (b) $\dfrac{5\pi r^3}{24}$ (c) $\dfrac{\pi r^2}{4}+\dfrac{2\pi r^3}{3}$ (d) $\dfrac{3\pi r^4}{8}$ **[2]**

Exam practice questions

3 The diagram shows a quarter of a circle, with radius 17 cm and centre O.
Points A and X lie on the circumference of the circle. The point N lies on OX and angle ANO = 90°.
ON = 8 cm and AN = 15 cm.

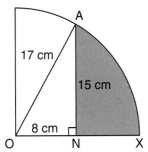

Find the shaded area on the diagram. **[5]**

4 In the diagram, OACB is a parallelogram. XY is parallel to OB. X is the mid-point of OA and N is the mid-point of XY.

$\overrightarrow{OA} = \mathbf{a}$ and $\overrightarrow{OB} = \mathbf{b}$

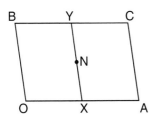

(a) Write, in terms of **a** and **b**:

$\overrightarrow{XN}$, $\overrightarrow{ON}$, $\overrightarrow{AN}$, $\overrightarrow{NB}$ **[3]**

(b) Deduce two facts about the points A, N, B. **[2]**

5 The diagram shows the position of three airports:

(A) Ayton, (B) Beesville and C (Colesville).

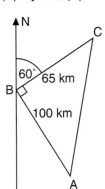

The distance from B to C is 65 km on a bearing of 060°,
angle CBA = 90° and AB = 100 km.

(a) Calculate, correct to three significant figures, the distance AC. **[3]**

(b) Calculate, to the nearest degree, the bearing of A from C. **[4]**

(c) An aircraft leaves B at 09.45 a.m. and flies directly to A arriving at 10.03 a.m. Calculate its average speed, giving your answer to an appropriate degree of accuracy. **[4]**

6 The diagram shows a cone with height 8 cm and base radius 6 cm. The curved surface of the cone is made from a sector of a circle which is also shown.

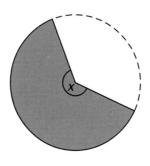

Find, leaving your answers as multiples of π:

(a) the area of the curved surface of the cone **[3]**

(b) the volume of the cone **[2]**

(c) the angle, x, made by the sector of the cone. **[3]**

7 The diagram shows a square-based pyramid constructed to advertise a chocolate product. The sloping edge length is 21.2 cm and the side of the base is 7.4 cm long.

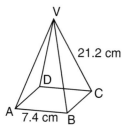

(a) Calculate the size of angle AVD. **[4]**

(b) The height of the pyramid. **[4]**

(c) The angle between edge VA and the base. **[3]**

8 (a) Sketch the curve $y = \cos x$ for values of x from 0 to 360°.
   Label the axes clearly. **[2]**

(b) Write down the coordinates of the points of intersection with the x-axis. **[2]**

Draw the line $y = 0.8$ on your sketch graph.
The x-coordinates of the points where the line meets the curve form the solution of an equation in x.

(c) Write down this equation. **[1]**

(d) Solve the equation. **[2]**

Unit 4 Handling Data

Overview

Topic	Section	Studied in class	Revised	Practice questions
4.1 Identification and selection	**Key questions**			
	Bias			
	Sampling			
	Primary and secondary data			
4.2 Collecting data	**Questionnaires and surveys**			
4.3 Processing and representing data	**Time series**			
	Moving averages			
	Scatter diagrams			
	Stem and leaf diagrams			
	Cumulative frequency			
	Quartiles			
	Histograms			
	Box and whisker plots			
	Finding the mean, median and mode of grouped data			
4.4 Probability	**Independent events**			
	Mutually exclusive events			
	Tree diagrams			
4.5 Interpretation	**Comparing data**			

4.1 Identification and selection

After studying this section, you will be able to:

● identify key questions to be asked and how to ask them
● recognise bias
● understand sampling and use sampling techniques
● distinguish between primary and secondary data

Key questions

AQA A   AQA B
EDEXCEL A   EDEXCEL B
OCR A   OCR B
OCR C
NICCEA
WJEC

As you will know from work completed earlier in your school life, handling data is concerned with the analysis and interpretation of data. The data is often collected in response to questions asked. As part of your coursework you will have to complete, or you may already have completed, a data handling task. This task may have involved the forming of a hypothesis. In order to do this you must ask questions and on the written examination papers you may be asked to make comments on the quality and appropriateness of questions to be asked as part of a survey.

Here are some 'issues' and some questions you might choose to ask:

(a) travel to school
    (i)   How long does it take you?
    (ii)   How far away from school do you live?
    (iii) How do you travel to school?

(b) holiday destinations
    (i)   Do you like to go somewhere hot?
    (ii)   When do you go?
    (iii) Do you like 'activity holidays'?

You might like to think of some more questions that it would be appropriate to ask for either of these issues.

However you must remember that the answers to the questions asked have to be analysed.

 KEY POINT

Those questions where the answer will be '*yes*' or '*no*' are clearly easy to analyse. Those where responses can be '*qualified*' such as '*do you eat breakfast*?' – answer '*sometimes*' and questions with an '*or*' statement in them, for example '*Do you eat meat or vegetables*?' should be avoided.

Bias

AQA A   AQA B
EDEXCEL A   EDEXCEL B
OCR A   OCR B
OCR C
NICCEA
WJEC

In data handling the word **population** is used for a collection, set or group of objects being studied.

Anything that distorts data so that it will not give a fairly representative picture of a population is called **bias**.

Two common ways in which bias arises are:

(a) through the style of the questions asked, e.g. asking a leading question such as '*Normal people watch the news on TV. Do you watch the news?*' and (b) from the population asked.

One way to avoid bias is by selecting an appropriate sample so that the results obtained represent the whole population. The size of a sample is important: it should be large enough to represent the population but small enough to be manageable.

Two examples of biased samples are:

(a) investigating the pattern of absence from a school by studying the registers in December. [This might be biased because (i) students are more likely to be ill in the winter months compared with, say, the summer months. (ii) Older students might be absent for interviews. (iii) The pattern of any truancy might vary.]

(b) finding out opinions about school dinners by asking the first 20 students in Year 11 seen in the dinner queue one day. [This might be biased because (i) only those eating school dinners might be asked – those eating sandwiches because they don't like school dinners might not be asked. (ii) The opinions of pupils in other years won't be recorded.]

Sampling

AQA A   AQA B
EDEXCEL A   EDEXCEL B
OCR A   OCR B
OCR C
NICCEA
WJEC

A **sample** is a small part of a population. Samples are used because it is quicker and cheaper to sample a population rather than to collect information from the whole population. Conclusions drawn from the samples can then be applied to the whole population.

If the structure or composition of the population is known then it is important to ensure that the sample (or samples) represent that population and thus any variations in the population should be reflected in the sample, which is therefore a representative sample.

Sampling methods

1 Systematic sampling

An example of this method is the selection of a 10% sample by going through the population picking every tenth item or individual. The drawback is that this only provides a representative sample if the population is arranged in a random way and not in a way that might introduce bias, for example if high or low values are grouped together.

Always show/state the type of sampling you are using and why.

2 Attribute sampling

In this method the selection of the sample is made by choosing some attribute which is totally unrelated to the variable being investigated. For example choosing a sample to investigate any relationship between head size and height from a list of people on the basis of their birthday being the first of the month.

3 Stratified sampling

The population is divided into strata or sub-groups and the sample chosen to reflect the properties of these sub-groups. For example, if the population contained three times as many people under 25 as over 25 then the sample should also contain three times as many people under 25. The sample should also be large enough for the results to be significant.

4 Random sampling

In random sampling every member of a population has an equal chance of being selected. The sample could be chosen by giving every member of the population a number and using random number tables, or the random number function on your calculator, to select the sample. To ensure a sample is random and as accurate as possible, ideally the sampling should always be repeated a number of times and the results averaged.

5 Quota sampling

This method is often used in market research where people interviewed have to be of a certain age, sex or social class etc.

6 Cluster sampling

The population is divided into small groups called clusters. One or more clusters are chosen using random sampling. This can lead to bias if the clusters are all different.

7 Stratified random sampling

A stratified random sample is obtained by:

- separating the population into appropriate categories or strata e.g. by age,
- finding out what proportion of the population is in each stratum,
- selecting a sample from each stratum in proportion to the stratum size.

This can be done by random sampling hence the technique is known as stratified random sampling.

Example

A survey about sport is carried out among students in Years 9, 10 and 11. There are 210 students in Year 9, 225 in Year 10 and 195 in Year 11. A sample of 50 students is taken.

The sample size from each of the three year groups must be in proportion to the stratum size so the 50 students are selected as follows:

There are $210 + 225 + 195 = 630$ students in total.

Year	Fraction of students	Number of students in the sample of 50
9	$\dfrac{210}{630}$	$\dfrac{210}{630} \times 50 = 17$
10	$\dfrac{225}{630}$	$\dfrac{225}{630} \times 50 = 18$
11	$\dfrac{195}{630}$	$\dfrac{195}{630} \times 50 = 15$

Round to the nearest student.

Select 17 students from Year 9, 18 from Year 10 and 15 from Year 11 at random.

A different form of sampling is illustrated by this example.

Example

The natterjack toad is an increasingly threatened species.

Scientists want to find out how many of the toads live in and around a pond.

To do this they catch 20 natterjack toads and mark them in a harmless way.

The toads are then released.

Next day another 20 are caught. 5 of these toads have already been marked, in other words 25% (5 out of a sample of 20) are already marked.

But 20 toads were marked initially.

This suggests that 25% of the population is about 20.

$$\frac{25}{100} \times P = 20$$

$$\therefore \quad P = 80$$

Therefore the total population is 80.

Primary and secondary data

Primary data are data collected by the person who is going to analyse and use it. **Secondary data** are data that are available from an external source such as books, newspapers, or the internet

Example

Which of the following are primary and which secondary data?

(a) Looking at records to see how many babies were born each day in December.

(b) Measuring the length of pebbles in a sample of pebbles from the beach.

(c) Counting the number of red cars passing the school gate.

(d) Phoning local shops, supermarkets, garages etc to find out how much the pay for a Saturday job is.

(e) Looking at the top ten charts to see which group is top each week.

(f) Finding out information for a holiday by looking at brochures.

(a) secondary   (b) primary   (c) primary   (d) primary

(e) secondary   (f) secondary

 KEY POINT   **Try to start with enough primary or secondary data that will allow you to sample from it.**

 PROGRESS CHECK

1 Why might this method give a biased sample?
An engineer is carrying out a traffic survey to find out how busy a particular road is. He counts the number of cars which pass a point on two days between 2p.m. and 3p.m.

2 For each of the following write down which sampling method is being used to carry out a survey of pupils in Years 10 and 11 in a school.

(a) Listing all the pupils in Year 10 and Year 11 in alphabetical order, then choosing the first and every fifth pupil after that.

(b) Numbering all the tutor groups from 1 to 10. Writing the numbers 1 to 10 on slips of paper and putting them into a bag which is shaken. One slip is taken out and the number on this slip gives the tutor group.

(c) Choosing the first 10 boys and the first 10 girls from the year groups who are in the dinner queue.

(d) Listing all the students in Year 10 and Year 11 and giving each a number and using random numbers to select the sample.

(e) There are 150 pupils in Year 10 and 165 pupils in Year 11. The sample contains 10 pupils from Year 10 and 11 from Year 11 chosen at random.

3 The table shows the numbers of boys and girls in Year 10 and Year 11 of a school.

	Year 10	Year 11
Boys	115	128
Girls	110	115

The headteacher wants to find out their views about changes to the school uniform and takes a stratified random sample of 40 pupils from Year 10 and Year 11.

Calculate the number of pupils to be sampled from Year 11.

PROGRESS CHECK

1 It may be biased because:
(a) the time is not appropriate – doesn't consider the rush hour for example (b) the length of time for the survey, 1 hour, is not sufficient (c) two days doesn't allow for variations i.e. could be Saturday, could be a holiday, could be the market day ...

2 (a) systematic (b) cluster (c) quota (d) random (e) stratified random

3 Year 11 = $\frac{243}{468} \times 40 = 21$ (to nr. whole number)

4.2 Collecting data

LEARNING SUMMARY

After studying this section, you will be able to:

● *write and use questionnaires*

● *make and use two-way tables*

Questionnaires and surveys

AQA A   AQA B
EDEXCEL A   EDEXCEL B
OCR A   OCR B
OCR C
NICCEA
WJEC

There are some standard ways of gathering information and data:

1 By questionnaire

● A questionnaire should give sufficient choices to cover the possible answers.

● The information must not be ambiguous.

● Answers should be short and capable of being analysed simply – yes/no types of responses are clearly the best.

● Questions should not be biased and should be short and easily understood.

● Questions should be relevant to the survey.

2 By observation

Here you need to consider:

- whether you are actually answering the question asked
- does the time and place of the observation matter?
- did you collect data for long enough?

3 By experiment

Questions to ask here include:

- does the experiment test the concept or hypothesis?
- have you carried out sufficient experiments producing enough results to reflect what is happening?

Questionnaire design

The best way to learn to write appropriate questions is to look at some good examples and some poor examples, and look back at the notes on bias in Section 4.1.

Remember, tick boxes are easy to analyse

Question	Comment
Normal students like pizza. Do you like pizza?	This is biased. The first sentence should not be there. It implies you aren't normal if you don't like pizza.
It is important to learn mathematics in school. Tick one box. ☐ Agree ☐ Disagree ☐ Don't know	This question is OK. It is clear and simple to analyse.
How old are you? ☐ 11–16 ☐ 16–18 ☐ 18–24	Where does a student aged 16 tick? The problem here is that the groups overlap, otherwise it would be a good question.

At the end of the survey you need to think about the presentation of your results. Decide whether to use scatter graphs, bar charts, histograms etc, which are discussed later, or two-way tables.

Two-way tables

AQA A   AQA B
EDEXCEL A   EDEXCEL B
OCR A   OCR B
OCR C
NICCEA
WJEC

These are used to show two sets of information.

Example

A teacher has conducted a survey of the students in Year 9 to find out their favourite language.

	French	German	Spanish	Total
Boys	25	45	20	90
Girls	50	43	53	146
Total	75	88	73	236

Reading this table shows that, for example, 25 boys preferred French, that 53 girls preferred Spanish, that 88 students preferred German.
Using tables like this will allow often you to find missing values.

Example

A group of 180 Year 7 and Year 8 students were asked if they preferred individual or team sports. 72 students out of a total of 96 Year 7 students said they preferred team sports and in total, 63 students preferred individual sports. The data is entered in a two-way table.

	Year 7	Year 8	Total
Individual sports			63
Team sports	72		
Total	96		180

The missing values can now be filled in giving:

	Year 7	Year 8	Total
Individual sports	24	39	63
Team sports	72	45	117
Total	96	84	180

PROGRESS CHECK

1 A bookseller records the type of book men and women buy during one week. Copy and complete the table.

	Men	Women	Total
hardback fiction		32	88
paperback fiction	74		274
hardback non-fiction	83	24	
paperback non-fiction		120	
Total	313		

	Men	Women	Total
hardback fiction	56	32	88
paperback fiction	74	200	274
hardback non-fiction	83	24	107
paperback non-fiction	100	120	220
Total	313	376	689

1

4.3 Processing and representing data

LEARNING SUMMARY

After studying this section, you will be able to:

- recognise and interpret time series
- calculate moving averages
- plot scatter graphs
- construct and interpret stem and leaf tables
- draw and interpret cumulative frequency curves, histograms and box plots
- calculate the mean, median and mode for grouped data

Time series

A time series is made up of numerical data recorded at intervals of time and plotted as a line graph. The diagram below shows some examples of time series:

Graph A shows random fluctuations.

Graph B shows regular fluctuations about the trend line: these are cyclical fluctuations.

Graph C shows seasonal fluctuations.

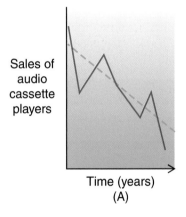

Time (years)
(A)

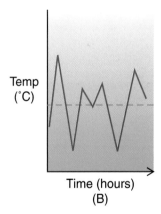

Time (hours)
(B)

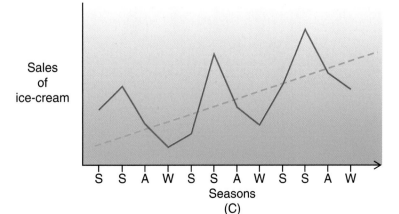

Seasons
(C)

Example

The diagram below shows the number of people who book their holidays with Lumsden's Travel Service.

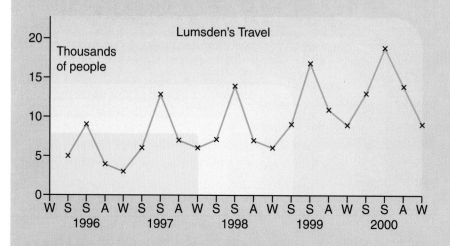

(a) How many people used the company in 1999?

(b) What was the percentage increase in people taking winter holidays between 1996 and 2000?

(a) 9000 + 17 000 + 11 000 + 9000 = 46 000

> Add up the Spring, Summer, Autumn and Winter numbers.

(b) $\frac{9000 - 3000}{3000} \times 100\% = 200\%$

> 9000 winter holidays in 2000, 3000 in 1996.

Moving averages

AQA A AQA B
EDEXCEL A EDEXCEL B
OCR A OCR B
OCR C
NICCEA
WJEC

Moving averages are those averages worked out for a given number of data items, as you work through the data. A four-point moving average uses four data items in each calculation, a three-point moving average uses three and so on.

Example

Find the three-point moving average for the following data:

10, 11, 6, 17, 16, 9, 14

For the first three values, 10, 11, 6 the average is (10 + 11 + 6) ÷ 3 = **9**. Miss out the first value, 10, and include the next value, 17, giving the three numbers 11, 6, 17. The average is (11 + 6 + 17) ÷ 3 = **11.3**. Carry on through the data values until you reach the point where you include the last value. The complete list of averages, is:

(10 + 11 + 6) ÷ 3 = **9**, (11 + 6 + 17) ÷ 3 = **11.3**, (6 + 17 + 16) ÷ 3 = **13**, (17 + 16 + 9) ÷ 3 = **14**, (16 + 9 + 14) ÷ 3 = **13**.

Moving averages can help you to draw a trend line on a time-series graph.

> **Example**
>
> Look back at the Lumsden's Travel Service graph. Here is the table showing the values used to draw the graph. The values of four-point moving averages are added.

Year	Quarter	Number of people, in 1000s	Four-point moving average
1996	Spring	5	
	Summer	9	
	Autumn	4	5.25
	Winter	3	5.5
1997	Spring	6	6.5
	Summer	13	7.25
	Autumn	7	8
	Winter	6	8.25
1998	Spring	7	8.5
	Summer	14	8.5
	Autumn	7	8.5
	Winter	6	9
1999	Spring	9	9.75
	Summer	17	10.75
	Autumn	11	11.5
	Winter	9	12.5
2000	Spring	13	13
	Summer	19	13.75
	Autumn	14	13.75
	Winter	9	

These values should be lined up between adjacent seasons eg 5.25 should appear *between* Summer and Autumn etc.

Plot these points on the graph and draw the trend line.

With data given in quarters like this, it is easy to plot the moving averages. After the first one they appear between the adjacent sections.

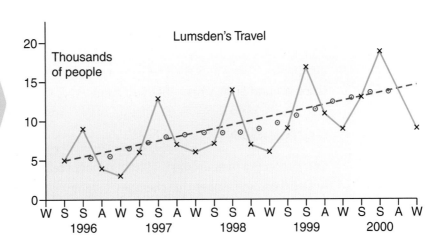

Lumsden's Travel

Scatter diagrams

AQA A
EDEXCEL A   AQA B
OCR A   EDEXCEL B
OCR C   OCR B
NICCEA
WJEC

Scatter diagrams, or scatter graphs, are used to see if there are any possible links or relationships between two features or variables. Values of the two features are plotted as points on a graph. If these points tend to lie in a straight line then there is a relationship or **correlation** between them.

> **Example**
>
> The table below shows the percentage marks gained in a history test and a geography test by the same 16 students.
>
> | History | 73 | 58 | 62 | 53 | 40 | 61 | 63 | 46 | 48 | 49 | 60 | 61 | 61 | 69 | 85 | 48 |
> |---|---|---|---|---|---|---|---|---|---|---|---|---|---|---|---|---|
> | Geography | 67 | 58 | 52 | 53 | 45 | 57 | 48 | 49 | 53 | 53 | 55 | 58 | 59 | 54 | 62 | 57 |
>
> This data is plotted on a scatter graph which suggests that there is a relationship between the results of the two subjects: the higher the history mark the higher the geography mark. The mean mark for each subject is calculated: history 58.6, geography 55.0, which gives the 'mean point', which is plotted. A straight line is drawn through this mean point and the middle of the other points, as shown.

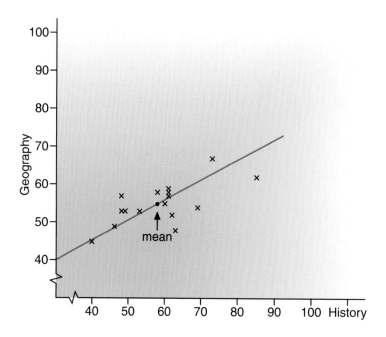

A seventeenth student took the geography test but missed the history test. The geography score was 50%. The scatter graph can be used to estimate the history score: around 50%. Check that you agree.

> **Draw lines across and up the graph to help.**

Correlation

Correlation is a measurement of how strong a relationship there is between two sets of data. Remember that there are different kinds of correlation.

Positive correlation

In this example as age increases so does height.

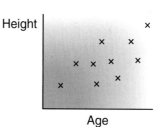

Negative correlation

In this example the value decreases as the age increases.

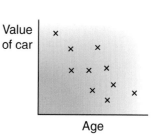

No correlation or zero correlation

In this example there is no correlation between maths score and height.

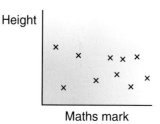

Stem and leaf tables

A stem and leaf table shows the shape of a distribution. It is similar to a bar chart with the numbers in the distribution forming the bars.

Example

Here are the marks gained by 30 students in a mathematics examination.

63	58	61	52	59	65	69	75	70	54
57	63	76	81	64	68	59	40	65	74
80	44	47	53	70	81	68	49	57	61

The marks can be shown in a **stem and leaf** table like this:

First: Divide each number into two parts: the tens and the units.

Then write the tens figures in the left-hand column of a table. These are the 'stems'.

Next go through the marks in turn and put in the units figure of each mark in the appropriate row. These are the 'leaves'.

the first value is 63	the next is 58	then 61
4 |	4 |	4 |
5 |	5 | 8	5 | 8
6 | 3	6 | 3	6 | 3 1
7 |	7 |	7 |
8 |	8 |	8 |

When all the marks are entered the table will look like this:

```
4 | 0 4 7 9
5 | 8 2 9 4 7 9 3 7
6 | 3 1 5 9 3 4 8 5 8 1
7 | 5 0 6 4 0
8 | 1 0 1
```

Finally rewrite the table so the units figures in each row are in size order, with the smallest first.

```
4 | 0 4 7 9
5 | 2 3 4 7 7 8 9 9
6 | 1 1 3 3 4 5 5 8 8 9
7 | 0 0 4 5 6
8 | 0 1 1
```

This is a finished stem and leaf table. It is a sort of frequency chart and allows you to to read off certain information, for example:

● the modal group, the one with the highest frequency, is the 60–69 group
● there are 30 results so the median result is mid-way between the fifteenth and the sixteenth results. Starting at the first result, 40, and counting on 15 results gives 63; the sixteenth result is also 63, so the median result is 63.

KEY POINT   **The stem is the first part of the number. The leaf has the end digits of the readings as in the example below.**

Example
The following table gives the lengths, in centimetres, of 20 worms in a sample.

4.4	5.3	4.9	5.8	5.2	5.9	6.2	6.4	6.5	6.0
6.3	6.4	7.6	7.2	7.6	8.1	9.3	9.2	7.3	7.7

(a)   Make a stem and leaf table to show the lengths.

(b)   Use your table to find the median length.

(a) The completed stem and leaf table is:

```
4 | 4 9
5 | 2 3 8 9
6 | 0 2 3 4 4 5
7 | 2 3 6 6 7
8 | 1
9 | 2 3
```

(b) There are 20 values. The median will be mid-way between the tenth
     and the eleventh values. The tenth value is 6.4 and the eleventh value is
     6.4 so the median is 6.4.

Cumulative frequency

In data handling the frequency tells you how often a particular result was
obtained. The **cumulative frequency** will tell you how often a result was
obtained which was less than, $<$, or less than or equal to, $\leqslant$, a given or stated
value in a collection of data.
The cumulative frequency is obtained by adding together the frequencies to give
a 'running total'.

For, **example,** as part of a statistics project a student collected information on
the numbers of brothers and sisters of children in Year 7. The results were
recorded in a cumulative frequency table.

Number of brothers and sisters	Frequency	Cumulative frequency	
0	40	40	
1	53	93	← $93 = 40 + 53$
2	37	130	← $130 = 93 + 37$
3	13	143	← $143 = 130 + 13$
4	6	149	← $149 = 143 + 6$
5	1	150	← $150 = 149 + 1$
6	1	151	← $151 = 150 + 1$

> **Note how the cumulative frequency is calculated.**

The median is the value half-way through the data. The cumulative total is 151
so the median is the seventy-sixth value. At the end of the '0s' you have reached
the fortieth value, at the end of the '1s' you have reached the ninety-third value
so the seventy-sixth value must be 1, i.e. the median is 1.

If there are a lot of values they are best dealt with by treating them as grouped
data.

Example

The table shows the marks gained in a test by the 60 pupils in a year group.

22	13	33	31	51	24	37	83	39	28
31	64	23	35	9	34	42	26	68	38
63	34	44	77	37	15	38	54	34	22
47	25	48	38	53	52	35	45	32	31
37	43	37	49	24	17	48	29	57	33
30	36	42	36	43	38	39	48	39	59

> 60 results are a lot to analyse so the results are grouped together in intervals.

Put the marks in a grouped frequency table.

Mark, m	Frequency	Cumulative frequency
$0 \leqslant m < 10$	1	1
$10 \leqslant m < 20$	3	4
$20 \leqslant m < 30$	9	13
$30 \leqslant m < 40$	25	38
$40 \leqslant m < 50$	11	49
$50 \leqslant m < 60$	6	55
$60 \leqslant m < 70$	3	58
$70 \leqslant m < 80$	1	59
$80 \leqslant m < 90$	1	60

> A sensible interval in this case is a band of 10 marks

The last column 'Cumulative frequency' gives the running total. Here the running total shows the number of pupils with less than a certain mark, for example 38 pupils gained less than 40 marks.

The `values for cumulative frequency can be plotted to give a cumulative frequency curve as shown below:

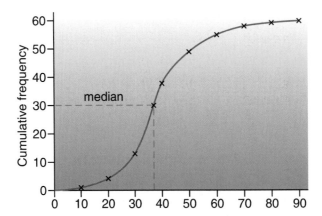

> Note that the cumulative frequency values are plotted at the upper value of each interval i.e. at 10, 20, 30 and so on because the intervals are based on discrete values i.e. marks.

> A cumulative frequency curve requires you to join the points with a curve. On a cumulative frequency graph you could join the points with straight lines.

You can use a cumulative frequency curve to estimate the median. In this example there are 60 pupils so the median mark will be the thirtieth mark. To find the median from the graph, find 30 on the vertical scale and go across the graph until you reach the curve and read off the value on the horizontal scale (look at the dotted line on the graph).

The median mark is about 37. Check that you agree.

Quartiles

AQA A AQA B
EDEXCEL A EDEXCEL B
OCR A OCR B
OCR C
NICCEA
WJEC

As the name suggests quartiles are associated with quarters.

For the example on page 147, 'The numbers of brothers and sisters in Year 7', one-quarter of the way through the data is the thirty-eighth value (the median value ÷ 2) which is 0, the upper quartile is three-quarters of the way through the data, i.e. 3 × 38 = 114, the 114th value which is 2.

The interquartile range is the difference between the lower quartile value and the upper quartile value: in this case 2 − 0 = 2

For the test marks example on page 148, the lower quartile is one-quarter of the way through the data and so is 60 ÷ 4, or the fifteenth value. Drawing a line across the graph at 15 gives a mark of 31. The upper quartile is three-quarters of the way through the data and so is 3 × 15, or the forty-fifth value. Drawing a line across at 45 gives a mark of 45.

The interquartile range is the difference between the lower quartile value and the upper quartile value: in this case 45 − 31 = 14.

Histograms

AQA A AQA B
EDEXCEL A EDEXCEL B
OCR A OCR B
OCR C
NICCEA
WJEC

Histograms and bar charts are closely related.
- In a histogram the frequency of the data is shown by the **area** of each bar. (In a bar chart the frequency is shown by the **height** of each bar.)
- Histograms have bars, or columns, whose width is in proportion to the size of the group of data each bar represents, the class width, so the bars may have different widths. (In a bar chart the widths of each bar are usually the same.)
- The vertical scale is labelled 'frequency density'. (In a bar chart the vertical scale is the actual frequency.)
- Histograms can only be used to show continuous data which is numerical and grouped.

Example
A botanist measured the height of a type of plant growing in compost. The table shows the measurements.

Height, h cm	Frequency
$0 \leqslant h < 10$	22
$10 \leqslant h < 20$	35
$20 \leqslant h < 30$	38
$30 \leqslant h < 40$	28
$40 \leqslant h < 50$	13
$50 \leqslant h < 60$	9

Noltice use of '⩽' so, for example $20 \leqslant h < 30$ means all values between 20 and 30, including 20 but excluding 30.

because the widths of all the groups, the class widths, are the same the histogram can be drawn without any further calculation.

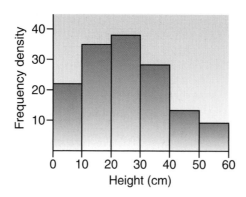

If the class widths are different then you must calculate the frequency densities as shown in the following example.

Example

A doctor investigated the ages of patients that visited her surgery over a period of time.
The table shows the findings.

Age, A, in years	Frequency
$0 \leqslant A < 20$	28
$20 \leqslant A < 30$	36
$30 \leqslant A < 40$	48
$40 \leqslant A < 50$	20
$50 \leqslant A < 70$	30
$70 \leqslant A < 100$	15

Here to draw the histogram you must first calculate the frequency densities:

Frequency density = frequency class ÷ class width

Class Age, A, in years	Frequency width	Frequency (f)	Frequency density
$0 \leqslant A < 20$	20	28	$28 \div 20 = 1.4$
$20 \leqslant A < 30$	10	36	$36 \div 10 = 3.6$
$30 \leqslant A < 40$	10	48	$48 \div 10 = 4.8$
$40 \leqslant A < 50$	10	20	$20 \div 10 = 2$
$50 \leqslant A < 70$	20	30	$30 \div 20 = 1.5$
$70 \leqslant A < 100$	30	15	$15 \div 30 = 0.5$

Note: frequency density = frequency per year

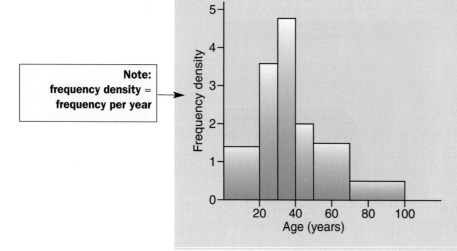

If the information is already given in a histogram it is possible to find out the frequencies and also estimate the mean.
This example shows how.

Example

A botanist measured the height of some bushes growing in a clay-type soil. The results are shown in this histogram.

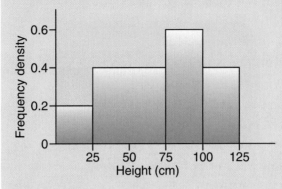

Find the number of bushes measured and estimate the mean height.

The frequency density = frequency ÷ column width.
The frequencies therefore are:
$0.2 \times 25 = 5$
$0.4 \times 50 = 20$
$0.6 \times 25 = 15$
$0.4 \times 25 = 10$
The total frequency = 50

frequency = frequency density × column width

The mean can be estimated using the mid-point of each class width.
$\Sigma fx = (5 \times 12.5) + (20 \times 50) + (15 \times 87.5) + (10 \times 112.5) = 3500$
Therefore $\bar{x} = \dfrac{3500}{50} = 70$ i.e. the mean height = 70 cm

KEY POINT Histograms often have no label on the vertical axis. It is assumed that the scale is frequency density.

Box and whisker plots

A box and whisker plot, sometimes just called a box plot, shows how the data is distributed. It can show the median, the quartiles and the range.

Example

A maths teacher recorded the times that a Year 7 class spent on their maths homework one night. The table shows the times, to the nearest minute, after they have been arranged in order, smallest to largest:

12	16	16	18	18	18	18	19	19	19
20	20	21	21	21	21	21	21	25	26
27	29	29	30	30					

The smallest value = 12, the largest = 30. The median is the thirteenth value = 21.
The lower quartile is the median of that data to the left of the actual median:

| 12 | 16 | 16 | 18 | 18 | 18 | 18 | 19 | 19 | 19 | 20 | 20 |

↑
lower quartile

The upper quartile is the median of that data to the right of the actual median:

| 21 | 21 | 21 | 21 | 21 | 25 | 26 | 27 | 29 | 29 | 30 | 30 |

↑
upper quartile

so the lower quartile = 18 and the upper quartile = 25.5.
You can now draw a box plot for the data.
the 'box' stretches from 18, the lower
quartile, to 25.5, the upper quartile. the
median is shown inside the box at 21, and
the 'whiskers' stretch from the lower quartile
to the smallest value, 12, and from the upper quartile to the highest value, 30.

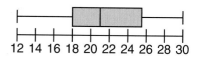

12 14 16 18 20 22 24 26 28 30

You can use a box plot to look at the shape of a distribution.

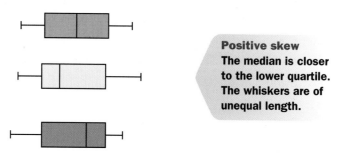

Symmetrical distribution
The median is in the centre.
The whiskers are of equal length.

Positive skew
The median is closer to the lower quartile.
The whiskers are of unequal length.

Negative skew
The median is closer to the upper quartile.
The whiskers are of unequal length.

Finding the mean, median and mode of grouped data

AQA A   AQA B
EDEXCEL A   EDEXCEL B
OCR A   OCR B
OCR C
NICCEA
WJEC

The mean, median and mode can be calculated from grouped data. This example shows how.

Example

Ann is completing a project comparing novels. She keeps a record of the number of words in each sentence in the first chapter of two novels. Here are her results for one novel.

number of words, w	frequency, f	cumulative frequency, cf	mid-value x	$f \times x$
1–5	16	16	3	$16 \times 3 = 48$
6–10	27	43	8	$27 \times 8 = 216$
11–15	28	71	13	$28 \times 13 = 364$
16–20	11	82	18	$11 \times 18 = 198$
21–25	12	94	23	$12 \times 23 = 276$
26–30	6	100	28	$6 \times 28 = 168$
31–35	5	105	33	$5 \times 33 = 165$
Totals	105			1435

Σf

Σfx

Modal class

Because the data is grouped the modal class is used instead of the mode. Here the **modal class** is 11–15 words as it has the highest frequency, 28.

Median

The median is estimated by 'interpolation'. There are 105 values so the middle value is the fifty-third value which will occur in the 11–15 group. This group is 5 numbers wide: 11, 12, 13, 14, 15. The frequency is 28 so this group has 28 values in it. The fifty-third value is the tenth of these 28 values because $53 = 43 + 10$, (the forty-third value is at the end of the 6–10 group) so an estimate would be $\frac{10}{28}$ of the way through this group.

So the median $= 11 + \frac{10}{28} \times 5$ giving 13 to the nearest whole number.

Mean

You cannot find the exact value of the mean when the data is grouped. You can estimate it by using the mid-value of each group. These are shown in the table as the x-values.
Multiply each x-value by the frequency, f, of the group, as shown in the last column of the table.
Add up the fx values to get Σfx.
Divide by the total frequency, Σf.

$$\text{Estimated mean} = \frac{\Sigma fx}{\Sigma f}$$
$$= \frac{1435}{105}$$
$$= 13.67$$
$$= 14 \text{ to the nearest whole number}$$

PROGRESS CHECK

1 The marks of 10 students in the two papers of a maths exam were:

Paper 1	20	32	40	45	60	67	71	80	85	91
Paper 2	15	25	40	40	50	60	64	75	76	84

Plot these marks on a scatter diagram.
A student scored 53 marks on paper 1. What would you estimate the likely mark to be on paper 2?

2 The table shows the length of time, in minutes, that customers queued at a supermarket checkout.

KEY POINT

Always explain why you are using mean, median or mode (modal class). Which is more appropriate depends on the context.

Time, t min	Number of customers
$0 < t \leqslant 1$	4
$1 < t \leqslant 2$	12
$2 < t \leqslant 3$	22
$3 < t \leqslant 4$	37
$4 < t \leqslant 5$	47
$5 < t \leqslant 6$	33
$6 < t \leqslant 7$	30
$7 < t \leqslant 8$	12

(a) Draw a cumulative frequency curve for the data.
(b) Estimate the median time.
(c) Estimate the upper and lower quartiles and hence find the interquartile range.

3 A clothing manufacturer needs to know how long to make the sleeves of sweatshirts. 200 teenagers had their arm lengths measured. The results are shown in the table:

Arm length, L cm	Frequency, f
$40 \leqslant L < 45$	8
$45 \leqslant L < 50$	44
$50 \leqslant L < 55$	96
$55 \leqslant L < 60$	28
$60 \leqslant L < 70$	20
$70 \leqslant L < 80$	4

Draw a histogram to show this information.

4 This histogram shows the ages of people who live in a small village.

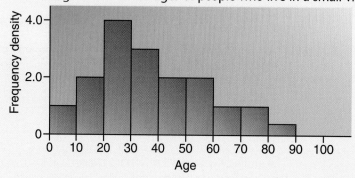

(a) How many people live in the village?
(b) Estimate their mean age.

5 John records how late his train is each working day over a 4-week period. These are the minutes late, to the nearest minute, for his journeys.

8	1	4	1	12	13	15	9	4	5
8	20	12	16	12	10	9	17	5	3

(a) Find the median.
(b) Find the upper and lower quartiles.
(c) Use these values to draw a box plot to show the times.

6 The mathematics marks of 250 students have been recorded and are given in the table below. Copy and complete the table and estimate the mean and the median examination marks.

Mark range	Mid-value, x	Frequency, f	$f \times x$
0–9		3	
10–19		4	
20–29		6	
30–39		19	
40–49		40	
50–59		49	
60–69		52	
70–79		43	
80–89		25	
90–99		9	

7 Mary has kept a record of the money she spends on food for her family at the supermarket each month for the last 12 months.
   (a) Draw a graph to show this.

Month	Jan	Feb	Mar	Apr	May	Jun	Jul	Aug	Sept	Oct	Nov	Dec
Amount, £	193	198	194	200	197	203	198	199	201	202	203	210

PROGRESS CHECK

   (b) Calculate a three point moving average for the data.
   (c) On your graph show the trend line.

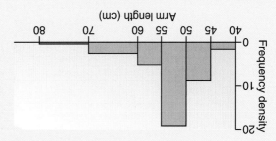

4.4 **Probability**

After studying this section, you will be able to:

- *understand what independent and mutually exclusive events are*
- *use the multiplication rule for independent events*
- *use the addition rule for mutually exclusive events*
- *draw and use tree diagrams*

Independent events

AQA A
AQA B
EDEXCEL A
EDEXCEL B
OCR A
OCR B
OCR C
NICCEA
WJEC

Two events are **independent** if the outcome of one event has no effect on the outcome of the other event. For example rolling a dice and spinning a coin will be independent events because whatever is scored on the dice can have no effect on whether a head or a tail is obtained on the coin.

Example

Jodi is using a spinner and a dice. She spins the spinner and rolls the dice.

What is the probability that she will get red on the spinner and a score less than 4 on the dice?

You could draw a table showing all the possible outcomes. This is called a **possibility space** diagram.

		Dice					
		1	2	3	4	5	6
	Red, R	R,1	R,2	R,3	R,4	R,5	R,6
Spinner	Green, G	G,1	G,2	G,3	G,4	G,5	G,6
	Yellow, Y	Y,1	Y,2	Y,3	Y,4	Y,5	Y,6
	Blue, B	B,1	B,2	B,3	B,4	B,5	B,6

The possible ways of gaining a red and a score of less than 4 are shown highlighted in the table.

The probability of getting red and less than $4 = \frac{3}{24} = \frac{1}{8}$.

However because the events in the above example are independent the probability can also be worked out using the **multiplication rule**.

If P(A) is the probability of gaining a red on the spinner, $P(A) = \frac{1}{4}$.

If P(B) is the probability of scoring less than 4 on the dice $P(B) = \frac{1}{2}$.

The probability of A and B happening is

The multiplication rule for independent events.

$$P(A \text{ and } B) = P(A) \times P(B) = \frac{1}{4} \times \frac{1}{2} = \frac{1}{8}.$$

Mutually exclusive events

AQA A AQA B
EDEXCEL A EDEXCEL B
OCR A OCR B
OCR C
NICCEA
WJEC

Events are **mutually exclusive** if they cannot happen at the same time. For example if you roll a dice it is impossible to get a 3 at the same time as getting an even number.

For two mutually exclusive events, A and B, the probability that either event A or event B will occur is found by **adding** their probabilities together.

The addition rule for mutually exclusive events.

$$P(A \text{ or } B) = P(A) + P(B)$$

Thus for the dice example above the probability of getting a 3 or getting an even number is

$$P(\text{getting a 3}) + P(\text{getting an even number}) = \frac{1}{6} + \frac{1}{2} = \frac{2}{3}.$$

Example
Asif and Rakhi have designed a game for the school fair. In the game an ordinary dice is rolled twice. A prize is won if the total score on both dice is 6 or if the score on each dice is over 4. What is the probability of winning a prize?

List all the possible results in a table:

This is a possibility space diagram.

1,1	1,2	1,3	1,4	1,5	1,6
2,1	2,2	2,3	2,4	2,5	2,6
3,1	3,2	3,3	3,4	3,5	3,6
4,1	4,2	4,3	4,4	4,5	4,6
5,1	5,2	5,3	5,4	5,5	5,6
6,1	6,2	6,3	6,4	6,5	6,6

Here the possible ways of scoring 6 are shown in blue, and the possible ways of each score being more than 4 are shown in red.

If P(A) is the probability of a total score of 6 then

$$P(A) = \frac{5}{36}$$

If P(B) is the probability of each score being more than 4 then

$$P(B) = \frac{4}{36}$$

The probability of a total of 6 or of each score being more than 4, i.e. P(A **or** B) can be seen from the table as $\frac{9}{36}$, and

$$P(A) + P(B) = \frac{5}{36} + \frac{4}{36} = \frac{9}{36}.$$

Tree diagrams

AQA A AQA B
EDEXCEL A EDEXCEL B
OCR A OCR B
OCR C
NICCEA
WJEC

Probability trees can be used to show the outcomes of two or more events. Each branch represents a possible outcome for an event. The probability of each outcome is written on the branch, the final result depends on the path taken through the tree.

Example

The probability that a new car will develop a fault in the first year is 0.85. Two new cars are chosen at random.

(a) Show all the possible outcomes on a tree diagram.

(b) Use the diagram to find the probability that both cars will develop a fault.

(c) Find the probability that only one car will develop a fault.

(a)
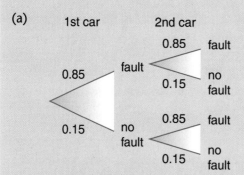

Remember that the sum of probabilities = 1. So if the probability of developing a fault = 0.85 the probability of not developing a fault = 0.15.

(b) The probability that both cars develop a fault = P(1st) × P(2nd)
= 0.85 × 0.85
= 0.7225

(c) The probability that only one car develops a fault means either the first car **or** the second car has a fault.

P(1st fault) × P(2nd no fault) + P(1st no fault) × P(2nd fault)

= (0.85 × 0.15) + (0.15 × 0.85)

= 0.255

So probability that only one car develops a fault is 0.255.

Example

A bag contains 7 red marbles and 4 white marbles. Two marbles are chosen without replacement (this means that the first marble is not put back in the bag before the second marble is chosen).

(a) What is the probability that both marbles are red?

(b) What is the probability that they are different colours?

(c) What is the probability that both marbles are the same colour?

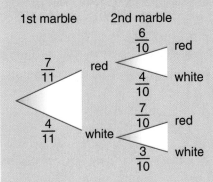

1st marble   2nd marble

(a) P (both red) = P(1st red and 2nd red) = $\dfrac{7}{11} \times \dfrac{6}{10} = \dfrac{21}{55}$

(b) P (one red and one white) = P((1st red and 2nd white) or (1st white and 2nd red))

$$= \left(\dfrac{7}{11} \times \dfrac{4}{10}\right) + \left(\dfrac{4}{11} \times \dfrac{7}{10}\right)$$

$$= \dfrac{14}{55} + \dfrac{14}{55} = \dfrac{28}{55}$$

(c) P (both the same) = P((1st red and 2nd red) or (1st white and 2nd white))

$$= \left(\dfrac{7}{11} \times \dfrac{6}{10}\right) + \left(\dfrac{4}{11} \times \dfrac{3}{10}\right)$$

$$= \dfrac{21}{55} + \dfrac{6}{55} = \dfrac{27}{55}$$

> **Note that the probability that both are the same = 1 – the probability that they are different**
>
> $= 1 - \dfrac{28}{55}$

PROGRESS CHECK

1 The probability that the school football team win their next match is 0.4. The probability that they draw their next match is 0.3. What is the probability that they win or draw their next match?

2 This spinner is used in a game. What is the probability that if it is spun twice you will get 'red' on both spins?

1 Mutually exclusive since cannot win and draw at the same time so
P(win or draw) = P(win) + P(draw) = 0.3 + 0.4 = 0.7

2 Independent events since second spin is not affected by first spin
p(red and red) = P(red) × P(red) = $\dfrac{1}{4} \times \dfrac{1}{4} = \dfrac{1}{16}$

4.5 Interpretation

After studying this section, you will be able to:

● analyse and compare data presented in a variety of forms

Comparing data

AQA A   AQA B
EDEXCEL A   EDEXCEL B
OCR A   OCR B
OCR C
NICCEA
WJEC

You need to be able to compare sets of data and draw conclusions about them. Here are some examples to illustrate the techniques.

Example

The children in a Year 7 class record the time it takes them to run between two marker posts.

The results are:

	Time (seconds)							
Boys	46	47	49	61	43	54	48	52
	55	55	50	49	48	54	49	46
Girls	46	48	48	56	47	54	55	64
	49	51	47	52	49	49	65	63

Compare the distributions of the times for boys and girls.

Boys		**Girls**	
mean = 50.38		mean = 52.68	
median = 49		median = 50	
mode = 49		mode = 49	
range = 18		range = 19	

For both boys and girls the median, mode and range are similar but the mean for the girls is slightly higher so one could deduce that the girls are slightly slower than the boys.

You could show the distributions using a stem and leaf table.

Here the numbers for girls and boys have been put in order.

```
            Girls                          Boys
  9   9   9   8   8   7   7   6 | 4 | 3   6   6   7   8   8   9   9   9
          6   5   4   2   1 | 5 | 0   2   4   4   5   5
              5   4   3 | 6 | 1
```

This is a 'back to back' stem and leaf table. The central column shows the 'tens' part of each number.

This gives an impression of the distribution of both sets of data and shows, for example, that more girls than boys took more than 60 seconds.

Example

Here are a list of English test results for two groups of students.

Group A	52	53	45	57	48	49	53	53	56	58	59	54	62	53
Group B	52	53	45	57	48	49	53	53	56	58	59			

Calculate the mean, median, mode and range for each set of data and compare the two groups' performance in English.

Group A   mean = 53.7, median = 53, mode = 53, range = 17
Group B   mean = 53, median = 53, mode = 53, range = 13

Comments could include: although the means are virtually the same, group B's is slightly lower than Group A's so one could argue that Group B is a little weaker in English but the range for Group B is smaller so the students in Group B are more consistent.

Example

A market gardener grows plum trees. He experiments with different fertilisers to see which fertiliser gives the best results. The table gives the weights, in grams, of the plums from 25 trees using each fertiliser.

Fertiliser A
79	91	48	86	44	67	54	36	83	55
79	26	82	68	77	80	18	42	61	76
24	20	56	73	69					

Fertiliser B
92	34	54	79	71	89	40	80	54	88
93	61	25	48	90	99	56	28	78	91
41	53	51	73	78					

Compare the two sets of data using the mean, median, mode and range. Which fertilser do you think is best?

Fertiliser A: mean = 59.8, median = 67, mode = 79, range = 73
Fertiliser B: mean = 62.9; median = 71; mode = 54 and 78, range = 74
Note the results for fertiliser B are bi-modal.
Fertiliser B seems to give better results for it shows a higher mean and median.

1 The temperatures of two towns were recorded for 12 days:

Day	1	2	3	4	5	6	7	8	9	10	11	12
Town A	11	13	12	11	14	15	17	15	16	15	20	18
Town B	10	12	15	13	16	12	15	16	14	16	17	21

Compare the temperatures for the two towns using mean, median, mode and range.

1 A: mean = 14.75, median = 15, mode = 15, range = 7
B: mean = 14.75; median = 15; mode = 16, range = 11
Town B has the higher mode, but the range of temperatures for town A is smaller so there is less variation.

Sample GCSE questions

1   In a survey conducted in a shopping centre shoppers were asked the distances, in kilometres, from their homes to the centre. The results are shown in the table:

Distance, x km	$0<x\leqslant5$	$5<x\leqslant10$	$10<x\leqslant15$	$15<x\leqslant20$	$20<x\leqslant25$	$25<x\leqslant30$
Frequency	38	42	76	30	10	4

(a)   Write down the modal class.   **[2]**

(b)   Calculate an estimate for the mean distance.   **[5]**

(c)   Complete the cumulative frequency table: **[3]**

Distance, x km	Cumulative frequency
$0<x\leqslant5$	38
$5<x\leqslant10$	
$10<x\leqslant15$	
$15<x\leqslant20$	
$20<x\leqslant25$	
$25<x\leqslant30$	

(d)   Draw a cumulative frequency graph, using a scale of 2 cm for 5 km on the horizontal axis and 1 cm for 10 shoppers on the vertical axis.   **[4]**

(e)   Use your graph to estimate:

(i)   the median distance   **[2]**

(ii)   the interquartile range   **[3]**

(f)   What percentage of shoppers live more than 6 km away from the centre?   **[4]**

The modal class is the class with the highest frequency

(a)   *The modal class is $10<x\leqslant15$.*   ✔✔

(b)   *Using the mid-points of each interval:*

$(38 \times 2.5) + (42 \times 7.5) + (76 \times 12.5) + (30 \times 17.5)$
$+ (10 \times 22.5) + (4 \times 27.5)$ ✔

$= 95 + 315 + 950 + 525 + 225 + 110$ ✔
$= 2220$ ✔

mean $= 2220 \div 200$ ✔
$= 11.1$ *km*   ✔

(c)

distance, x km	cumulative frequency
$0<x\leqslant5$	38
$5<x\leqslant10$	80
$10<x\leqslant15$	156
$15<x\leqslant20$	186
$20<x\leqslant25$	196
$25<x\leqslant30$	200

✔✔✔
(Lose 1 mark for each error)

Sample GCSE questions

(d)

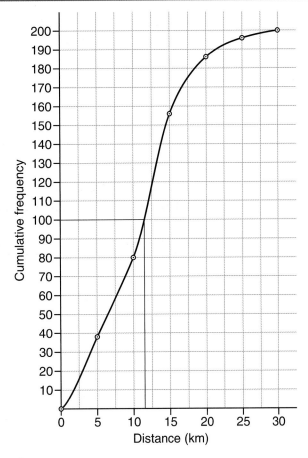

Distance (km)

✓✓✓✓

> *Plot the points at the end of each class interval, i.e. at 5, 10, 15, etc.*

(1 mark for correct axes, scales and labels
2 marks for correctly plotting points, but lose 1 mark for each mistake.
1 mark for a correct line drawn through the points.)

(e) (i) *median distance = 11.5 km* ✓✓
(this mark would be awarded on correctly interpreting your graph, so draw line across the page at the median value to show how you obtained your answer).

(ii) *lower quartile = 6.75, upper quartile = 14.25, interquartile range = 7.5* ✓✓✓

(f) *number of shoppers ⩽ 6 km = 44* ✓
number ⩾ 6 km = 156 ✓

$$percentage = \frac{156}{200} \times 100 = 78$$ ✓✓

2 The letters of the word 'PROCESSES' are written on individual cards and placed in a bag.

| P | R | O | C | E | S | S | E | S |

A letter is selected at random from the bag and then replaced.

(a) What is the probability that the letter is:

(i) E [1]

(ii) a vowel [1]

Sample GCSE questions

(iii) T   [1]

(iv) not S   [1]

The experiment is repeated 300 times.

(b) How many times would you expect the letter S to be chosen?   [3]

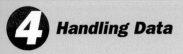

> This is an easy probability question included as revision.

(a) (i) $\dfrac{2}{9}$ ✔

(ii) $\dfrac{3}{9} = \dfrac{1}{3}$ ✔

(iii) 0   ✔

(iv) $\dfrac{6}{9} = \dfrac{2}{3}$ ✔

(b) *Probability of choosing S* $= \dfrac{1}{3}$ ✔

> Always simplify the fractions where possible

number of times S is chosen in 300 trials $= \dfrac{1}{3} \times 300 = 100$ ✔✔

3 Julie is taking a French examination. It consists of three parts: a reading test, a writing test and a talking and listening test.
The probability of her passing the reading test is 0.6. The probability of her passing the writing test is 0.7. The probability of her passing the talking and listening test is 0.4.

(a) Complete the tree diagram showing the possible outcomes of the three tests taken in the order shown.   [5]

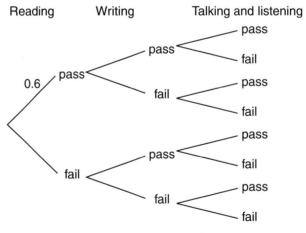

(b) What is the probability that she passes all three tests?   [3]

(c) What is the probability that she passes at least two of the tests?   [4]

(d) If Julie only passes one test she needs to repeat all three tests later in the year. What is the probability that Julie repeats all three tests?   [2]

Sample GCSE questions

(a)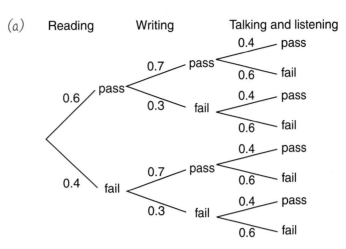

Reading   Writing   Talking and listening

(You will score 1 mark for completing the reading section, 2 marks for completing the writing section and 2 marks for completing the talking section.)

(b) Multiplying along the branches:
$0.6 \times 0.7 \times 0.4$ ✔✔
$= 0.168$ ✔

(c) The probabilities are added together giving:
$(PPP) + (PPF) + (PFP) + (FPP)$ ✔
$= 0.168 + 0.252 + 0.072 + 0.112$ ✔✔
$= 0.604$ ✔

(d) 1 subtract the answer to part (c) $= 1 - 0.604$ ✔
$= 0.396$ ✔

4 A dentist's patients are divided by age into groups as shown in the table below.

These questions usually involve independent and exclusive events. Make sure you know when to multiply and when to add.

Age, x years	Number of patients
$0 < x \leqslant 5$	14
$5 < x \leqslant 15$	41
$15 < x \leqslant 25$	59
$25 < x \leqslant 45$	70
$45 < x \leqslant 75$	16

(a) Draw a histogram to show this distribution. [5]

(b) The dentist wishes to choose a stratified random sample of 40 patients.
Show how this can be done. [4]

Sample GCSE questions

Calculate the frequency densities.

(a)

Age, x (years)	Number of patients	Frequency density
$0 \leqslant x < 5$	14	2.8
$5 \leqslant x < 15$	41	4.1
$15 \leqslant x < 25$	59	5.9
$25 \leqslant x < 45$	70	3.5
$45 \leqslant x < 75$	16	0.5

✔✔✔

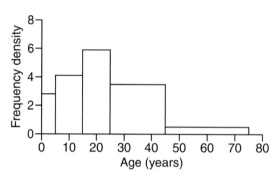

✔✔

(b) The total number of patients $= 14 + 41 + 59 + 70 + 16 = 200$ ✔
the numbers required in each age group are therefore:

$\dfrac{14}{200} \times 40 = 2.8$, i.e. 3 ✔✔✔

$\dfrac{41}{200} \times 40 = 8.2$, i.e. 8

$\dfrac{59}{200} \times 40 = 11.8$, i.e. 12

$\dfrac{70}{200} \times 40 = 14$

$\dfrac{16}{200} \times 40 = 3.2$, i.e. 3

Show your calculations so that you will earn method marks even if you make a mistake.

(3 marks for these calculations, lose 1 for each mistake)

Exam practice questions

1 The graph shows the quarterly sales, in £10 000s of plants and garden equipment at Shrubs'R'Us garden centre over a four-year period.

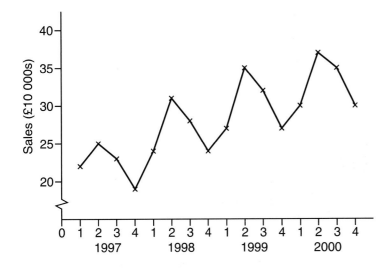

Year	Quarter	Sales £10 000's	Four-point moving average
1997	1	22	
	2	25	
	3	23	
	4	19	
1998	1	24	
	2	31	
	3	28	
	4	24	
1999	1	27	
	2	35	
	3	32	
	4	27	
2000	1	30	
	2	37	
	3	35	
	4	30	

(a) Copy the table and calculate the four-point moving averages for this data and enter the values on the table. **[3]**

(b) Plot these moving averages on the graph. **[2]**

(c) Draw the trend line. **[1]**

Exam practice questions

2 Martin's Bank have conducted a survey of people with savings accounts with the bank. The results are shown in the table:

Age, y (years)	Number of people	Cumulative frequency
$0 \leqslant y < 5$	0	
$5 \leqslant y < 15$	100	
$15 \leqslant y < 25$	500	
$25 \leqslant y < 35$	800	
$35 \leqslant y < 45$	1000	
$45 \leqslant y < 55$	1400	
$55 \leqslant y < 65$	2500	
$65 \leqslant y < 75$	900	

(a) Complete the cumulative frequency table.   **[2]**
(b) Draw a cumulative frequency curve for this data.   **[1]**
(c) Use your graph to estimate:
    (i)   the interquartile range of the ages   **[2]**
    (ii) the number of people over 62 years old who have savings accounts with the bank.
                                                                                                 [2]

3 The table gives the distances travelled to work each day by a group of 50 employees:

Distance, d (miles)	Frequency
$0 < d \leqslant 2.5$	16
$2.5 < d \leqslant 3.5$	9
$3.5 < d \leqslant 4.5$	11
$4.5 < d \leqslant 6.5$	10
$6.5 < d \leqslant 10.5$	4

(a) Fred claims that the median and mode of these distances are both equal to 4.6 miles. Which of these claims is incorrect and why?   **[2]**
(b) Construct a histogram to display the data.   **[3]**

4 (a) Jodi has 10 T-shirts. 7 are white and 3 are green.
       She chooses one at random.
       Naomi has 8 T-shirts. 1 is white, 2 are green and 5 are red.
       She chooses one at random.
       What is the probability that Jodi and Naomi choose the same coloured T-shirt?
                                                                                                 [5]

Exam practice questions

(b) Jen has 7 T-shirts. 2 are white, 4 are green and 1 is red. She chooses three at random.
What is the probability that Jen chooses T-shirts that are all the same colour?

[3]

5 A hospital recorded the birth weights, in kilograms, of 100 boys and 100 girls. The weights are summarised in the table below.

	Minimum	Lower quartile	Median	Upper quartile	Maximum
Girls	1.5	2.5	3.2	3.8	4.6
Boys	1.3	2.3	3.4	4.1	4.9

(a) Draw two box plots for the weights of the boys and girls to show this data. **[3]**
(b) Compare the weights of the boys and the girls. **[2]**

6 A coin is biased. The probability that it lands showing heads is 0.7.
To play a game the coin is tossed three times.
Find the probability that more heads than tails are obtained. **[3]**

Exam practice answers

Unit 1

1 (a) $2 \times 3 \times 3 \times 5 \times 13$ or $2 \times 3^2 \times 5 \times 13$

(b) (i) $\dfrac{1}{2} \times 4 = 2$

(ii) $\dfrac{\sqrt{3}}{3\sqrt{3}} = \dfrac{1}{3}$

(c) $3 \times 10^{-2} + 8 \times 10^{-1} = 0.83 = 8.3 \times 10^{-1}$

In part (a) there are part marks for a correct method or some correct factors, e.g., if 13 was omitted but $2 \times 3^2 \times 5$ shown, then 2 marks would be given. Intermediate steps or their equivalents earn part marks in (b) and (c), as shown. Correct answers earn full marks but to omit the working risks scoring zero.

2 (a) (i) $1\dfrac{15}{100} = 1\dfrac{3}{20}$

(ii) If F is the value of the fraction, then,
$$100F = 115.1\dot{5}$$
$$F = 1.1\dot{5}$$
Subtracting, $\quad 99F = 114$
$$F = \dfrac{114}{99} = 1\dfrac{15}{99} = 1\dfrac{5}{33}$$

(b) (i) $1 + 2 \times 1 \times \sqrt{2} + (\sqrt{2})^2 = 3 + 2\sqrt{2}$
Irrational

(ii) $1 + \sqrt{2} - \sqrt{2} - (\sqrt{2})^2 = 1 - 2 = -1$
Rational

(iii) $\dfrac{1 - \sqrt{2}}{(1 - \sqrt{2})(1 + \sqrt{2})}$
$$= \dfrac{1 - \sqrt{2}}{-1} = \sqrt{2} - 1$$
Irrational

For recurring decimals, always multiply by the power of 10 that lines up the pattern. To demonstrate that a number is rational or irrational it is necessary to get the number in the form shown in these answers. In part (b)(iii) the surd is in the denominator and the trick is to multiply top and bottom by a number that makes the denominator rational. Notice that part (ii) helped to find this number.

3 (a) $7.99 \div 1.175 = 68$p

(b) (i) $\dfrac{79.9}{69.9} \times 100 = 114.3$

14.3% increase

(ii) $79.9 \times (1.143)^5 = 155.9$p

Don't be tempted to multiply by 0.825 in part (a) – this is a reverse percentage. A good check is to increase your answer by 17.5% to see if it goes back to 79.9. There is a mark in (b)(ii) for multiplying by any power of 1.143.

4 (a) Maximum distance $= 2.05 \times 92.5$
$$= 189.625 \text{ km}$$

(b) Least fuel rate $= \dfrac{22.5}{189.625}$
$$= 0.12 \text{ litres/km}$$

In each case there is a mark for choosing correctly the highest or lowest value. Don't forget to find the least value, divide by the greatest.

5 (a) $1\dfrac{2}{5} + 3\dfrac{3}{4} = 4 + \dfrac{2}{5} + \dfrac{3}{4}$
$$= 4 + \dfrac{8 + 15}{20} = 4\dfrac{23}{20} = 5\dfrac{3}{20}$$

(b) $1\dfrac{2}{5} \times 3\dfrac{3}{4} = \dfrac{7}{5} \times \dfrac{15}{4}$
$$= \dfrac{7}{1} \times \dfrac{3}{4} = \dfrac{21}{4} = 5\dfrac{1}{4}$$

(c) $3\dfrac{1}{3} - 2\dfrac{5}{6} = \dfrac{10}{3} - \dfrac{17}{6} = \dfrac{20 - 17}{6}$
$$= \dfrac{3}{6} = \dfrac{1}{2}$$

The solution to part (b) shows the common factor 5 cancelled but you could multiply out first to get $(\dfrac{105}{20})$ and then cancel. This is longer. There is a short way in part (c). $3\dfrac{1}{3}$ is $\dfrac{1}{3}$ more than 3 and $2\dfrac{5}{6}$ is $\dfrac{1}{6}$ less than 3. $\dfrac{1}{3} + \dfrac{1}{6} = \dfrac{1}{2}$.

Unit 2

1 (a) (i) $6 - 3x = 2x - 12$
$$18 = 5x$$
$$x = 3\dfrac{3}{5}$$

(ii) $3x - 2(x - 1) = 6$
$$x + 2 = 6$$
$$x = 4$$

(b) (i) $4x \leqslant 6x - 9$
$$9 \leqslant 2x$$
$$x \geqslant 4\dfrac{1}{2}$$

(ii) $x \leqslant 15$
$$x \geqslant -15$$
(or $-15 \leqslant x \leqslant 15$)

Take care with the direction of the inequality signs in part (b)(i). The method shown (keeping the x- term positive and reading the inequality backwards) avoids dividing by a negative number. However, this gives the same solution ($-2x \leqslant -9, x \geqslant 4\frac{1}{2}$, changing the inequality sign on dividing by -2).

2 (a)
$$\frac{1}{v} = \frac{1}{u} - \frac{1}{f}$$
$$\frac{1}{v} = \frac{f - u}{uf}$$
$$v = \frac{uf}{f - u}$$

(b)
$$p(r - q) = rq$$
$$pr - pq = rq$$
$$rq + pq = pr$$
$$q(r + p) = pr$$
$$q = \frac{pr}{r + p}$$

In part (b), don't miss the fact that the new subject occurs twice in the formula and so you must first collect those terms.

3 x pence is the cost of a cup of coffee, y pence is the cost of a biscuit.
$$2x + 3y = 323 \qquad 3x + y = 292$$
$$2x + 3y = 323$$
$$9x + 3y = 876$$
$$-7x = -553$$
$$x = 79\text{p}$$
$$y = 292 - 3 \times 79 = 55\text{p}$$

Choose your own letters for the variables. If you do not use an algebraic method to find the solution, you will lose most of the marks.

4 (a)

x	-1	-0.5	0	0.5	1	1.5	2	3
y	-2	0.375	1	0.625	0	-0.125	1	10

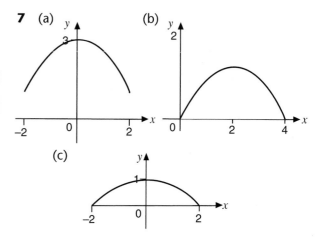

Draw a smooth curve through the points plotted.

(b) (i) $x = -0.6, 1, 1.6$
(ii) $x = 2.6$

(c) $x = 2.6 \qquad f(x) = 0.056$
$ x = 2.55 \qquad f(x) = -0.423...$
$ x = 2.59 \qquad f(x) = -0.042...$
$ x = 2.595 \qquad f(x) = 0.0067...$
 Solution $x = 2.59$

The solutions in part (b)(i) should be those from your graph. If you have made a mistake in drawing the graph, these marks could be earned so long as the readings were taken at the appropriate y values. To obtain full marks in part (c), you should show that the solution is nearer 2.59 than 2.6.

5 $y = 3 - 2x$
$$y^2 = 9 - 12x + 4x^2$$
$$x^2 + 9 - 12x + 4x^2 = 16$$
$$5x^2 - 12x - 7 = 0$$
$$x = \frac{12 \pm \sqrt{12^2 + 4 \times 5 \times 7}}{2 \times 5}$$
$$x = 2.89 \quad \text{or} \quad -0.49$$
$$y = -2.77 \quad \text{or} \quad 3.97$$
Points are $(2.89, -2.77)$, $(-0.49, 3.97)$

6 (a) $(x - 5)(x + 3)$
(b) $x = 5$ or $x = -3$
(c) Either $x - 5 > 0$ and $x + 3 > 0$ so $x > 5$
or $x - 5 < 0$ and $x + 3 < 0$ so $x < -3$

In part (c), the brackets must both have the same sign, i.e., both negative or both positive, if the product is to be positive. This gives pairs of inequalities. both of which must be satisfied. However, only one counts, for example, if $x > 5$ then it is certainly > -3 but there are values for which $x > -3$ but not > 5, so $x. > 5$ is the required condition. Similarly for the other pair.

7 (a) (b)

(c)

To check your work, test to see if known points fit. For example, in part (b), when $x = 2$, $y = f(0)$. From the original sketch of the function, $f(0) = 2$. Your sketch should have gone through $(2, 2)$.

Unit 3

1 (a) (i) 105° (corresponding angles) [1]
  (ii) 25° (= 180 − 50° − 105°,
       angle on a straight line) [1]
  (iii) 25° (alternate angles) [1]
  (iv) 130° (angles in an isosceles triangle) [2]

 (b) Angles PQX and SXR are equal
      (corresponding angles)
   Angles QXP and XRS are equal
      (corresponding angles)
   QX = XR (given) [1]
   Therefore triangles are congruent
      (ASA) [1]

   In part (b) the question says 'prove' so
   explanations and reasons are required.

2 (b) Could be volume because it is the only
   formula with just one term involving r^3.
   The $\dfrac{\pi r^2}{4}$ in the answer to part (c) means
   that this answer represents something which
   is not just volume.

3 $\cos \text{AON} = \dfrac{8}{17}$
   $\therefore$ angle AON = 61.9° [1]
   Area of sector OAX $= \dfrac{61.9}{360} \times \pi \times 17^2 = 156 \text{ cm}^2$ [1]
   AN $= \sqrt{17^2 - 8^2} = 15$ [1]
   Area of triangle AON $= \dfrac{1}{2} \times 8 \times 15 = 60 \text{ cm}^2$ [1]
   $\therefore$ shaded area
   $= 156 - 60 = 96 \text{ cm}^2$ [1]

   Follow through marks would be available in this
   multi-step problem. This means that if you made
   a mistake in your calculation in an early part of
   the solution as long as you used your wrong
   answer correctly later on no further loss of marks
   would arise.

4 (a) $\overrightarrow{XN} = \dfrac{1}{2}\overrightarrow{XY}$ and $\overrightarrow{XY} = \overrightarrow{OB} = \mathbf{b}$

   $\therefore \ \overrightarrow{XN} = \dfrac{1}{2}\mathbf{b}$ [1]

   $\overrightarrow{ON} = \overrightarrow{OX} + \overrightarrow{XN} \ = \dfrac{1}{2}\mathbf{a} + \dfrac{1}{2}\mathbf{b}$ [1]

   $\overrightarrow{AN} = \overrightarrow{AX} + \overrightarrow{XN} \ = -\dfrac{1}{2}\mathbf{a} + \dfrac{1}{2}\mathbf{b} = \dfrac{1}{2}\mathbf{b} - \dfrac{1}{2}\mathbf{a}$ [1]

   $\overrightarrow{NB} = \overrightarrow{NY} + \overrightarrow{YB} \ = \dfrac{1}{2}\mathbf{b} - \dfrac{1}{2}\mathbf{a}$ [1]

 (b) $\overrightarrow{AB} = \overrightarrow{AC} + \overrightarrow{CB} \ = \mathbf{b} - \mathbf{a}$ [1]
   $\overrightarrow{AN} = \dfrac{1}{2}(\mathbf{b} - \mathbf{a})$
   $\overrightarrow{AB} = \dfrac{1}{2}(\mathbf{b} - \mathbf{a})$

 $\therefore$ A, N, B all lie on the same straight line,
 with N as mid-point. [1]
 Take care over the directions and hence the signs
 of the vectors.

5 (a) AC^2 = 100^2 + 65^2 [1]
   AC = 119.27 km [1]
    = 119 km (to 3 s.f.) [1]

 (b) $\tan \text{BCA} = \dfrac{100}{65}$ [1]
   Angle BCA = 56.98° ie 57° [1]
   Bearing = 180° + 3°
    = 183° [2]

 (c) Speed $= \left(\dfrac{100 \text{ km}}{18 \text{ min}}\right) = \dfrac{100 \times 60}{18}$ [2]
    = 333 km/h (to 3 s.f.) [2]

 In part (b) draw a 'north-
 south' line at C as follows:

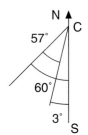

6 (a) (i) Curved surface area $= \pi r l$
       $l = \sqrt{8^2 + 6^2}$
        = 10 cm [1]
     Curved surface area $= \pi \times 6 \times 10$
        $= 60\pi \text{ cm}^2$ [2]

   (ii) $V = \dfrac{1}{3}\pi r^2 h$

     $= \dfrac{1}{3} \times \pi \times 36 \times 8$ [1]
     $= 96\pi \text{ cm}^3$ [1]

 (b) Area of sector $= \dfrac{x}{360} \times \pi r^2$ [1]

   $\therefore \ \dfrac{x}{360} \times \pi 10^2 = 60\pi$ [1]

   $\therefore \ x = \dfrac{60 \times 360}{100} = 216°$ [1]

 As with question 3 there would be follow
 through marks available here. Remember to
 show all your working through the intermediate
 steps. Remember too that the radius of the sector
 is the siant height of the cone. (Make a cone
 from a sector of a circle like this if you are not
 sure.)

7 (a)

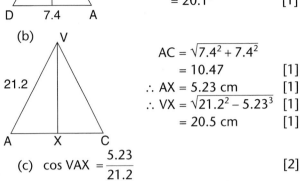

Angle AVD = $2 \times \sin\left(\dfrac{3.7}{21.2}\right)$ [2]

= 2×10.05 [1]

= $20.1°$ [1]

(b)

$AC = \sqrt{7.4^2 + 7.4^2}$

= 10.47 [1]

∴ AX = 5.23 cm   [1]

∴ VX = $\sqrt{21.2^2 - 5.23^3}$ [1]

= 20.5 cm   [1]

(c) $\cos VAX = \dfrac{5.23}{21.2}$ [2]

∴ angle VAX = $75.7°$ [1]

Draw sketches, as shown above, to help you identify sides and angles. Again follow through marks would be available - but remember to show all your working.

8 (a) This is a standard drawing, see page 96   [2]

(b) (90°, 0), (270°, 0)   [2]

(c) $\cos x = 0.8$ [1]

(d) If $\cos x = 0.8$ there are 2 solutions so the angle could be in the first or the fourth quadrant – see page 95
Solving $x = \cos^{-1} 0.8$ on a calculator gives x = 36.87° ie 36.9°   [1]
∴ the other angle = 360° – 36.9° = 323.1°   [1]

Draw a sketch to help you decide where the second solution lies:

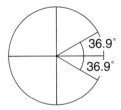

36.9°
36.9°

Unit 4

1 (a) Averages are:
(in £10 000s) 22.25, 22.75, 24.25, 25.5, 26.75, 27.5, 28.5, 29.5, 30.25, 31, 31.5, 32.25, 33

(b) Check your graph, plotting the values given in part (a).   [2]

(c) You should be able to draw a straight line passing through virtually every point.   [1]

Remember to plot the moving averages between the quarterly values given. On the graph the first average is plotted at the point (2.5, 22.25), the next at (3.5, 22.75) and so on.

2 (a) Cumulative frequencies are 0, 100, 600, 1400, 2400, 3800, 6300, 7200

(b) You can check the accuracy of your curve by looking at the given answers for part (c).

(c) (i) UQ = 61, LQ = 39, IQR = 22   [1 + 1]
(ii) 1600 to 1700   [1]

In part (c) remember to find the median first and then to find the LQ and the UQ. An easy way of doing this is to treat the lower quartile as the median of the values below the actual median and the upper quartile as the median of the values in the upper half, i.e. above the actual median.

3 (a) Median cannot be 4.6 because 36 distances are less than this.

(b) Frequency densities are 5, 10, 14, 4.5, 1.75.

For part (b) you are not given the scales to use so you must choose sensible scales for both axes, for example use 2 cm for 2 miles on the horizontal axis and 2 cm for 5 units on the frequency density, (the vertical) axis.

4 (a) Probability that Jodi chooses white = $\dfrac{7}{10}$ [1]
and that she chooses green = $\dfrac{3}{10}$
Probability that Naomi chooses white = $\dfrac{1}{8}$ [1]
and that she chooses green = $\dfrac{2}{8}$
Therefore the probability that both choose
white = $\dfrac{7}{10} \times \dfrac{1}{8} = \dfrac{7}{80}$ [1]
probability that both choose
green = $\dfrac{3}{10} \times \dfrac{2}{8} = \dfrac{6}{80}$ [1]
Therefore the probability that both choose
the same
= P(W,W) + P(G,G) = $\dfrac{7}{80} + \dfrac{6}{80} = \dfrac{13}{80}$ [1]

(b) To pick 3 of the same colour Jen must pick green.   [1]
First 'pick' the probability = $\dfrac{4}{7}$; for the second
the probability = $\dfrac{3}{6}$, and, for the third, the
probability = $\dfrac{2}{5}$ [1]
∴ the probability = $\dfrac{4}{7} \times \dfrac{3}{6} \times \dfrac{2}{5} = \dfrac{4}{35}$ [1]

Remember (i) when to add probabilities
(ii) when to multiply
(iii) that the sum of probabilities is 1

5 (a) Check your drawings using the values given in the table. there would be 1 mark for the girls' box plot correct, 1 mark for the boys' and 1 mark for using suitable scales   [3]

(b) Boys' average weight is greater than girls'. Boys' weights are more spread out, therefore more varied. The interquartile range, i.e. the middle 50% is wider than for the girls.

[1 mark for any valid observation up to a maximum of 2]

Remember to comment about the position of the median leading to a comment about spread and distribution and try to compare rather than comment on only the boys or the girls.

6

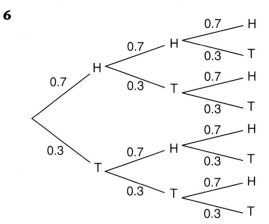
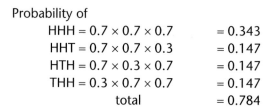

Probability of

HHH = $0.7 \times 0.7 \times 0.7$		= 0.343
HHT = $0.7 \times 0.7 \times 0.3$		= 0.147
HTH = $0.7 \times 0.3 \times 0.7$		= 0.147
THH = $0.3 \times 0.7 \times 0.7$		= 0.147
	total	= 0.784

[1 mark awarded for multiplying probabilities, 1 mark for adding the separate totals and 1 mark for the final answer]

A probability tree diagram, which may be given in the question (although you would have to complete by marking on the probabilities) is a logical method to use in questions like this because it shows all the possibilities.

Centre number	
Candidate number	
Surname and initials	

 **Examining Group**

General Certificate of Secondary Education

Mathematics
Higher Tier
Paper 1

Time: two hours

Instructions to candidates

Do **not** use a calculator.

Write your name, centre number and candidate number in the boxes at the top of this page.

Answer ALL questions in the spaces provided on the question paper.

Show all stages in any calculations and state the units.

Include diagrams in your answers where this may be helpful.

Information for candidates

The number of marks available is given in brackets **[2]** at the end of each question or part question.

The marks allocated and the spaces provided for your answers are a good indication of the length of answer required.

For Examiner's use only	
1	
2	
3	
4	
5	
6	
7	
8	
9	
10	
11	
12	
13	
14	
15	
16	
17	
18	
19	
20	
Total	

EDUCATIONAL

Useful Formulae

Volume of prism = (area of cross section) × length

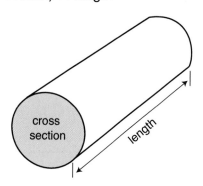

In any triangle *ABC*

Sine rule $\dfrac{a}{\sin A} = \dfrac{b}{\sin B} = \dfrac{c}{\sin C}$

Cosine rule $a^2 = b^2 + c^2 - 2bc\,\cos A$

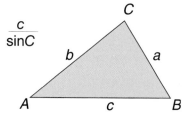

Area of triangle $= \dfrac{1}{2}ab\,\sin C$

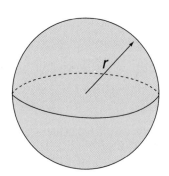

Volume of sphere $= \dfrac{4}{3}\pi r^3$

Surface area of sphere $= 4\pi r^2$

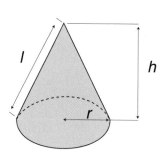

Volume of cone $= \dfrac{1}{3}\pi r^2 h$

Curved area of cone $= \pi r l$

The quadratic equation

The solutions of $ax^2 + bx + c = 0$, where a ≠ 0, are given by $x = \dfrac{-b \pm \sqrt{(b^2 - 4ac)}}{2a}$

Letts

1 Work out an estimate for the value of:

$$\frac{62.5 \times 5.07 - 9.89 \times 3.06}{97.8^2}$$

Give your answer as a fraction in its simplest form.

....

....

....

....

.... **[3]**

(Total 3 marks)

2 Here are the first five terms of an arithmetic sequence.

7, 10, 13, 16, 19

Find an expression in terms of n, for the nth term of the sequence.

....

.... **[2]**

(Total 2 marks)

3 Using the information that:

$28 \times 321 = 8988$

Write down the value of:

(i) 2.8×321 ....

.... **[1]**

(ii) 28×0.321 ....

.... **[1]**

(iii) $89.88 \div 2.8$ ....

.... **[1]**

(Total 3 marks)

Leave blank

Letts

4   On the grid triangle B is the image of triangle A after reflection.

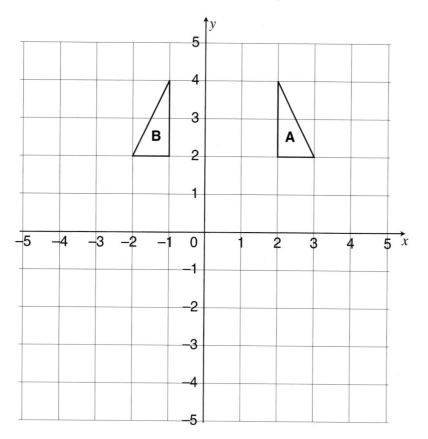

(a) Write down the equation of the line of reflection.

.... **[1]**

(b) Rotate triangle A through 90° clockwise about (2,1)

[2]

(c) Translate triangle B by the vector $\begin{pmatrix} -2 \\ -4 \end{pmatrix}$

[2]
(Total 5 marks)

5 Solve $6a - 3 = 3(a - 5)$

....

....

a .... **[2]**
(Total 2 marks)

6 The table shows some expressions.

The letters a, b and c represent lengths.

Place a tick in the appropriate column for each expression to show whether the expression can be used to represent a length, an area, a volume or none of these.

Expression	Length	Area	Volume	None of these
$a^2 + bc$				
$\dfrac{abc}{\pi b^2}$				
$3a^2 \sqrt{b^2 + c^2}$				

[3]
(Total 3 marks)

7 The times in minutes taken by 11 people to wait to be served at a supermarket checkout are listed in order:

1, 1, 2, 3, 3, 3, 4, 5, 5, 6, 8

(a) Find:

(i) the lower quartile

....

.... minutes

(ii) the interquartile range

....

.... minutes **[3]**

(b) Draw a box plot for this data on the grid below.

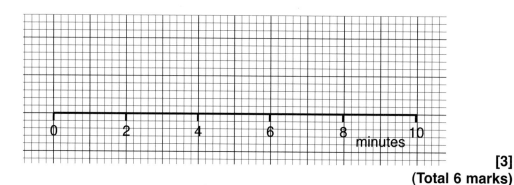

[3]

(Total 6 marks)

8 $a = 6 \times 10^7$

$b = 3 \times 10^4$

(a) Find the value of $a \times b$

Give your answer in standard form.

....

.... **[2]**

(b) Find the value of $\frac{a^2}{b}$

Give your answer in standard form.

....

....

.... **[3]**

(Total 5 marks)

9 **(a)** Simplify:

(i) $\dfrac{n^6}{n^2}$ .... **[1]**

(ii) $\dfrac{2h^4 \times 3h^7}{9h^{12}}$ .... **[1]**

(b) Expand and simplify:

(i) $(5x - 3)(2x + 7)$

.... **[2]**

(ii) $(5x - 4)^2$

.... **[3]**

(c) Solve the equation

$x^2 + 4x - 12 = 0$

.... **[3]**

(Total 10 marks)

10

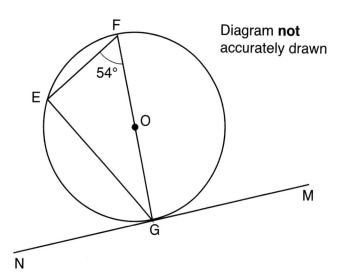

Diagram **not** accurately drawn

In the diagram E, F and G are points on the circle, centre O.

Angle EFG is 54°.

MN is a tangent to the circle at point G.

(a) Calculate the size of angle FGE.
Give reasons for your answer.

....

....

.... °   **[2]**

(b) Calculate the size of angle EGN.
Give reasons for your answer.

....

....

.... °   **[2]**

(Total 4 marks)

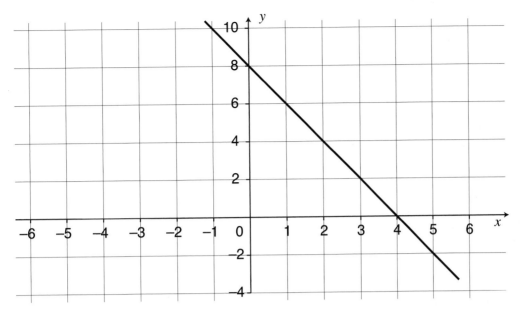

A straight line has been drawn on the grid.

(a) Write down the equation of the straight line.

....

....

$$y = \quad \text{....................}$$ **[2]**

(b) Write down the gradient of the line $x + 3y = 6$.

....

.... **[2]**

(c) Write down the equation of the line which is parallel to the line with the equation $4x + 2y = 8$ and passes through the point with coordinates $(0, -1)$.

....

....

.... **[2]**

(d) Write down the equation of a line which is perpendicular to the line $y = 3x$.

....

.... **[2]**

(Total 8 marks)

12 Work out:

(a) 5^0 .... **[1]**

(b) 8^{-2} .... **[1]**

(c) $27^{\frac{2}{3}}$ .... **[1]**

(d) $\frac{1}{25}^{-\frac{1}{2}}$ .... **[2]**

(Total 5 marks)

13 Rearrange the formula $y = \dfrac{a(x + b)}{x - c}$

to make x the subject.

....

....

....

....

.... **[4]**

(Total 4 marks)

Leave blank

14 A bag contains 4 red beads, 2 black beads and 3 green beads.
Rosie takes a bead at random from the bag, records its colour and replaces it.
She does this two more times.

Work out the probability that of the three beads Rosie takes out at least two are the same colour.

.... **[6]**
(Total 6 marks)

15 Work out:

$$\frac{(7 + \sqrt{5})\ (7 - \sqrt{5})}{\sqrt{80}}$$

Leave your answer in the form $\frac{a\sqrt{b}}{c}$.

....

....

....

.... **[4]**
(Total 4 marks)

16 a is directly proportional to the square of b.

When $a = 12$, $b = 2$.

  (a) Find an expression for a in terms of b.

....

....

$a = $ .... **[3]**

  (b) Calculate a when $b = 3$.

....

....

.... **[1]**

  (c) Calculate b when $a = 192$.

....

....

.... **[2]**

(Total 6 marks)

17 The table and histogram gives information about how long in minutes, some students took to complete a puzzle.

Time (t) in minutes	Frequency
$0 < t \leqslant 5$	....
$5 < t \leqslant 15$	52
$15 < t \leqslant 30$	48
$30 < t \leqslant 50$	88
$50 < t \leqslant 60$	....

Letts

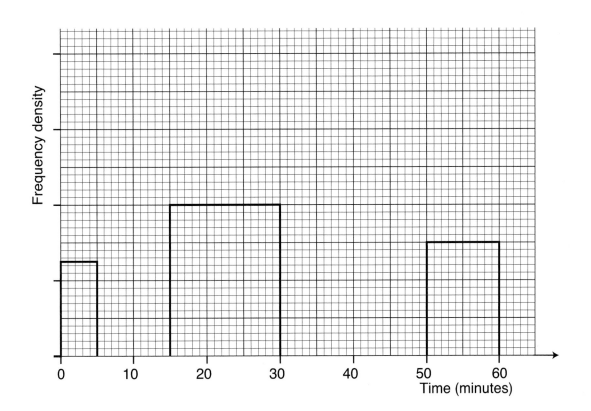

Frequency density

Time (minutes)

(a) Use the information in the histogram to complete the table. **[2]**

(b) Use the table to complete the histogram. **[2]**

(Total 4 marks)

18 The diagram shows a toy.
The toy is made up of a cone and a hemisphere.
Work out the volume of the toy.
Leave your answer in terms of π.

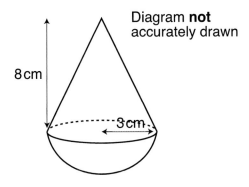

Diagram **not**
accurately drawn

8 cm

3 cm

.... **[7]**

(Total 7 marks)

19 Simplify fully:

$$\frac{2n^2 + n - 6}{4n^2 - 9} \times \frac{4n + 6}{n^2 + 3n + 2}$$

[6]
(Total 6 marks)

20 In this diagram $\overrightarrow{OD} = \mathbf{d}$, $\overrightarrow{OC} = 2\mathbf{c}$ and $\overrightarrow{OE} = 3\mathbf{d}$

F is the midpoint of CD and CG = $\frac{1}{4}$ CE.

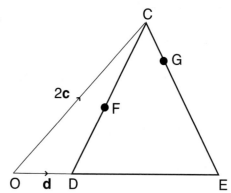

(a) Express in terms of **c** and **d**.

   (i) $\overrightarrow{DC}$ .... [1]

   (ii) $\overrightarrow{DF}$ .... [1]

(b) Prove that O, F and G lie on a straight line.

....

....

.... [5]
(Total 7 marks)

Centre number	
Candidate number	
Surname and initials	

Letts Examining Group

General Certificate of Secondary Education

Mathematics
Higher Tier
Paper 2

Time: two hours

Instructions to candidates

You **are expected to** use a calculator.

Write your name, centre number and candidate number in the boxes at the top of this page.

Answer ALL questions in the spaces provided on the question paper.

Show all stages in any calculations and state the units.

Include diagrams in your answers where this may be helpful.

Information for candidates

The number of marks available is given in brackets **[2]** at the end of each question or part question.

The marks allocated and the spaces provided for your answers are a good indication of the length of answer required.

You may refer to the formulae on page 2 of Paper 1.

For Examiner's use only	
1	
2	
3	
4	
5	
6	
7	
8	
9	
10	
11	
12	
13	
14	
15	
16	
17	
18	
19	
Total	

EDUCATIONAL

1 Calculate:

$$\frac{2.76 \times 3.27^2 - 2.93 \cos 30°}{\sqrt{12.62} + 0.327}$$

Give your answer to **four** significant figures.

....

.... **[3]**

(Total 3 marks)

2 p is an integer such that $-4 \leqslant 4p < 14$.

(a) List all the possible values of p.

.... **[2]**

(b) Solve the inequality:

$$\frac{3t + 2}{4} < t - 3$$

.... **[3]**

(Total 5 marks)

3 **(a)** Express the following numbers as products of their prime factors.

(i) 40

.... **[2]**

(ii) 105

.... **[2]**

(b) Find the Highest Common Factor of 40 and 105.

.... **[1]**

(c) Work out the Lowest Common Multiple of 40 and 105.

.... **[2]**

(d) Change the decimal $0.3\dot{6}$ into a fraction in its lowest terms.

.... **[3]**

(Total 10 marks)

Letts

4 The diagram represents a triangular playground DEF.
The scale of the diagram is: 1cm represents 2m.
A slide is to be placed in the playground so that it is:

12m from point **F**
equidistant between **D** and **E**.

On the diagram, mark the point with a letter **S** where the slide can be placed.

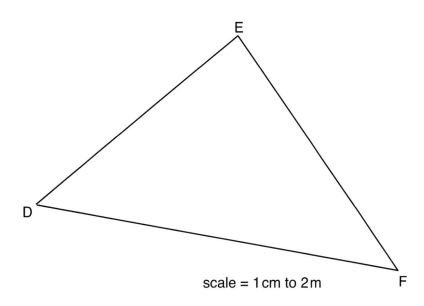

scale = 1cm to 2m

[3]
(Total 3 marks)

5 Use the method of trial and improvement to solve the equation:

$t^3 + 2t = 39$

Give your answer to one decimal place.
You must show all your working.

....

....

....

t .... **[4]**
(Total 4 marks)

6 The grouped frequency table shows information about the number of hours spent travelling by each of 60 commuters in one week.

Number of hours spent travelling (t)	Frequency
$0 < t \leqslant 5$	0
$5 < t \leqslant 10$	14
$10 < t \leqslant 15$	21
$15 < t \leqslant 20$	15
$20 < t \leqslant 25$	7
$25 < t \leqslant 30$	3

(a) Find the class interval in which the median lies.

....

.... **[2]**

(b) Work out an estimate for the mean number of hours spent travelling by commuters that week.

....

....

....

.... **[4]**

(Total 6 marks)

7

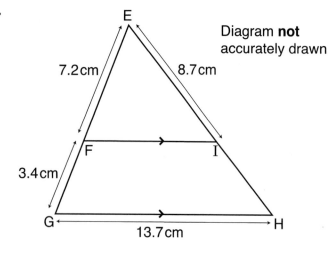

Diagram **not** accurately drawn

FI is parallel to GH

EF = 7.2 cm, FG = 3.4 cm

EI = 8.7 cm, GH = 13.7 cm

(a) Calculate the length of IH.

....

....

....

.... cm   **[3]**

(b) Calculate the length of FI.

....

....

....

.... cm   **[2]**

(Total 5 marks)

8 $p^3 = a^2b$

$a = 4 \times 10^6$

$b = 2 \times 10^3$

Find p.

Give your answer in standard form correct to 3 significant figures.

....

....

....

$p =$ ....   **[3]**

(Total 3 marks)

Letts

9

Diagram **not**
accurately drawn

C

9 cm

55°

A   B

The diagram shows triangle ABC.

BC = 9 cm
Angle ABC = 90°
Angle CAB = 55°

Work out the perimeter of the triangle.
Give your answer correct to 3 significant figures.

....

....

....

....

.... cm   **[5]**
(Total 5 marks)

10 A shop sells a television set.
It offers a discount of 15% off the normal price.
Ahmed buys the television set for £357.
Calculate the normal price of the television set.

....

....

....

£ ....   **[3]**
(Total 3 marks)

	Year 7	Year 10
Boys	90	50
Girls	85	70

The table shows the number of boys and the number of girls in year 7 and year 10.
The deputy head wants to find out how much homework pupils have per week.
A stratified sample of size 50 is to be taken from year 7 and year 10.

(a) Calculate the number of pupils to be sampled from year 7.

....

....

....

.... **[2]**

(b) Two pupils are to be chosen at random to speak to the headteacher.
One pupil is to be chosen from year 7.
One pupil is to be chosen from year 10.

Calculate the probability that a girl and a boy will be chosen.

....

....

....

.... **[3]**
(Total 5 marks)

12 This is the graph of $y = 2x^2 - 3x + 2$.

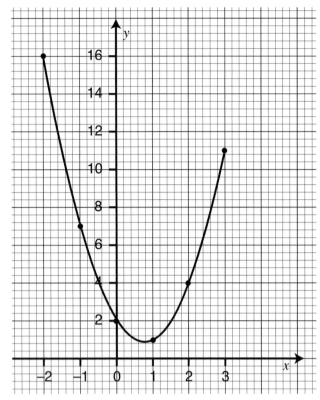

Use the graph to solve the following equations.

(a) $2x^2 - 3x + 2 = 2$

.... **[2]**

(b) $2x^2 - 3x + 2 = 10$

.... **[2]**

(c) $2x^2 - 4x - 2 = 0$

.... **[4]**
(Total 8 marks)

Letts

13 The length of a rectangle is 6.2 cm, correct to 1 decimal place.
The width of the rectangle is 3.76 cm correct to 2 decimal places.

(a) Calculate the upper bound for the area of the rectangle.
Write down all the figures on your calculator display.

....

....

....

.... cm^2 **[3]**

(b) $a = 7.46$ cm, correct to 2 decimal places.
$b = 6.3$ cm, correct to 1 decimal place.

Calculate the lower bound for $\dfrac{a^2}{b}$.

Write down all the figures on your calculator display.

....

....

....

.... cm   **[3]**
(Total 6 marks)

14

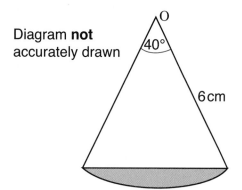

Diagram **not**
accurately drawn

The diagram shows a sector of a circle centre O.
The radius of the circle is 6 cm.
The angle at the centre of the circle is 40°.

Calculate the area of shaded segment.
Give your answer correct to 3 decimal places.

....

....

....

....

....

....

.... cm^2 **[6]**
(Total 6 marks)

15

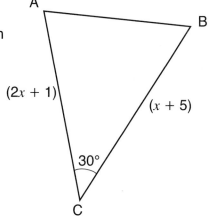

Diagram **not**
accurately drawn

The diagram shows a triangle.
The measurements of the diagram are in centimetres.
The lengths of the two sides are $(2x + 1)$ cm and $(x + 5)$ cm.
Angle ACB is 30°.
The area of the triangle is 10 cm^2.

(a) Show that:

$$2x^2 + 11x - 35 = 0$$

....

....

....

....

.... **[4]**

(b) Find the value of x.
Give your answer correct to 2 decimal places.

....

....

....

....

.... **[3]**

(Total 7 marks)

Letts

16 The population of a country is increasing at an annual rate of 1.3%.
In 1992, it was 42 million.
Calculate an estimate for the population of this country in 2012.

....

....

....

.... million   **[3]**

(Total 3 marks)

17 Solve the simultaneous equations:

$y = 2 - 3x$
$x^2 + y^2 = 58$

....

....

....

....   **[6]**

(Total 6 marks)

18 The diagram shows a quadrilateral ABCD.

Diagram **not**
accurately drawn

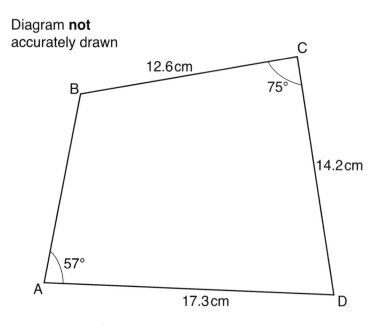

AD = 17.3 cm, BC = 12.6 cm, CD = 14.2 cm
Angle BCD = 75°, Angle BAD = 57°.

Calculate the size of angle ABD.
Give your answer correct to 2 decimal places.

..............° **[6]**
(Total 6 marks)

19 **(a)** Prove algebraically that the sum of the squares of any two consecutive integers is an odd number.

[3]

(b) Show that $(2n + 1)^2 - (n + 2)^2 = 3(n - 1)(n + 1)$

[3]
(Total 6 marks)

Question	Answer				Mark

1

$$\frac{60 \times 5 - 10 \times 3}{100^2}$$

$$= \frac{270}{10000}$$

$$= \frac{27}{1000} \quad \text{or} \quad \frac{3}{100}$$

3

Examiner's tip
Round each value to 1 significant figure.

2 $3n + 4$ 2

3   **i**   898.8   1
    ii   8.988   1
    iii   32.1   1

4   **a**   Reflection in the line $x = \dfrac{1}{2}$ 1
    b   2
    c   2

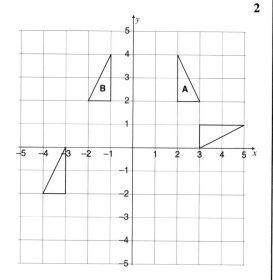

5   **a** $a = -4$ 2

Examiner's tip
Remember to multiply out the brackets first and then collect like terms.

6 $a^2 + bc$ Area

    $\dfrac{abc}{\pi b^2}$ Length

    $3a^2\sqrt{b^2 + c^2}$ Volume   3

7   **a**   **i**   lower quartile: 2 minutes
      ii   interquartile range: 3 minutes   3

    b

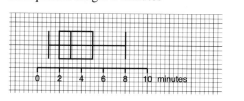

3

Examiner's tip
Remember that each section of the box plot represents 25%.

8   **a** 1.8×10^{12} 2

    b

$$\frac{(6 \times 10^7)^2}{3 \times 10^4} = \frac{36 \times 10^{14}}{3 \times 10^4}$$

$$= 12 \times 10^{10}$$

$$= 1.2 \times 10^{11}$$

3

Examiner's tip
You can use the law of indices on these questions.
Remember to write your final answer in standard form.

9   **a**   **i** n^4 1
      ii $\dfrac{2}{3h}$ 1

    b   **i** $10x^2 + 29x - 21$ 2
      ii $25x^2 - 40x + 16$ 3

    c $x = 2$ or $x = -6$ 3

Examiner's tip
In (b)(ii) remember that $(5x - 4)^2 = (5x - 4)(5x - 4)$. In part (c) when solving the quadratic equation, factorise first.

10   **a**   36°
      Since angle FEG = 90°, angles in a semicircle, then angle FGE = 180° − 90° − 54° since angles in a triangle add up to 180°.   2

    b   54°
      Since angle NGO = 90° because a radius and tangent meet at 90° and FGE = 36°, and EGN = 54°.   2

Examiner's tip
When asked to 'explain' make sure you refer to the circle theorems and not just show working out.

11 a $y = -2x + 8$ 2
    b $-\dfrac{1}{3}$ 2

c $y = -2x - 1$ **2**

d $y = \dfrac{-1}{3}\,x +$ any constant   **2**

Examiner's tip

For part (d) remember that two lines are perpendicular if the gradients multiply to give –1 ($\dfrac{-1}{3} \times 3 = -1$).

---

12   **a**   1   **1**

     b $\dfrac{1}{8^2} = \dfrac{1}{64}$ **1**

     c $(\sqrt[3]{27})^2 = 9$ **1**

     d $\dfrac{1}{25^{-\frac{1}{2}}} = 25^{\frac{1}{2}} = \sqrt{25} = \pm5$ **2**

13 $y = \dfrac{a(x + b)}{x - c}$

     $y(x - c) = ax + ab$
     $yx - yc = ax + ab$
     $yx - ax = yc + ab$
     $x(y - a) = yc + ab$

     $x = \dfrac{cy + ab}{(y - a)}$ **4**

Examiner's tip

Remember to show each step in your working.

---

14 $\dfrac{144}{729} = \dfrac{16}{81}$

     $1 - \dfrac{16}{81} = \dfrac{65}{81}$ **6**

Examiner's tip

You could draw a tree diagram to help you or work out the probability of drawing all three colours (i.e. red, green, blue) and mulitply by 6 as there are six possible combinations, then subtract this from 1.

15

$$\frac{(7 + \sqrt{5})(7 - \sqrt{5})}{\sqrt{80}}$$

$$= \frac{(49 - 5)}{\sqrt{80}}$$

$$= \frac{44}{\sqrt{16} \times \sqrt{5}}$$

$$= \frac{44}{4\sqrt{5}}$$

$$= \frac{11}{\sqrt{5}} \times \frac{\sqrt{5}}{\sqrt{5}}$$

$$= \frac{11\sqrt{5}}{5}$$

    4

Examiner's tip

To rationalise the denominator, multiply both the numerator and the denominator by.

---

16 a $a = 3b^2$ **3**

   b   27   **1**

   c ±8 **2**

17 a

Time (t) in minutes	Frequency
$0 < t \leqslant 5$	10
$5 < t \leqslant 15$	52
$15 < t \leqslant 30$	48
$30 < t \leqslant 50$	88
$50 < t \leqslant 60$	24

    2

b

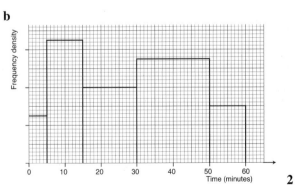

    2

Examiner's tip

Use the $15 < t \leqslant 30$ column to find that 1cm^2 represents a frequency of 4.

Question	Answer	Mark

18

Cone: $\frac{1}{3}\pi r^2 h$

$= \frac{1}{3} \times \pi \times 9 \times 8$

$= 24\pi$

Hemisphere: $\frac{2}{3}\pi r^3$

$= \frac{2\pi \times 3^3}{3}$

$= 18\pi$

Volume $= 42\pi\,\text{cm}^3$ **7**

Examiner's tip
One mark is for stating the units (cm^3).

19

$$\frac{2n^2 + n - 6}{4n^2 - 9} \times \frac{4n + 6}{n^2 + 3n + 2}$$

$$= \frac{(2n - 3)(n + 2)}{(2n - 3)(2n + 3)} \times \frac{2(2n + 3)}{(n + 1)(n + 2)}$$

$$= \frac{2}{(n + 1)}$$ **6**

Question	Answer	Mark

Examiner's tip
Factorise each part first and then cancel.

20 a i $\overrightarrow{DC} = -\mathbf{d} + 2\mathbf{c}$ **1**

     ii $\overrightarrow{DF} = \frac{1}{2}(-\mathbf{d} + 2\mathbf{c})$ **1**

    b $\overrightarrow{OF} = \frac{1}{2}(\mathbf{d} + 2\mathbf{c})$

      $\overrightarrow{OG} = \frac{3}{4}(\mathbf{d} + 2\mathbf{c})$ **5**

Since $\overrightarrow{OG}$ is a multiple of $\overrightarrow{OF}$
i.e. $\overrightarrow{OG} = \frac{3}{2}\overrightarrow{OF}$, and since they go
through the common point O; O, G and F must
lie on a straight line.

Examiner's tip
Be very systematic when working out vectors.

Total: 100 marks

Answers: GCSE Maths paper 2

Question	Answer	Mark

1   7.497 (4 s.f.)   **3**

Examiner's tip
Check your answer.

2 a $-1, 0, 1, 2, 3$ **2**

   b $t > 14$ **3**

Examiner's tip
Solve inequalities in a similar way to equations. Watch the
direction of the inequality sign.

3 a i $2 \times 2 \times 2 \times 5$ **2**

     ii $3 \times 5 \times 7$ **2**

   b   HCF $= 5$ **1**

   c   LCM $= 840$ **2**

   d $\frac{36}{99} = \frac{4}{11}$ **3**

Examiner's tip
When dividing to find factors, be systematic, e.g. start
with 2. When no more factors of 2, try 3 and so on.

Question	Answer	Mark

4

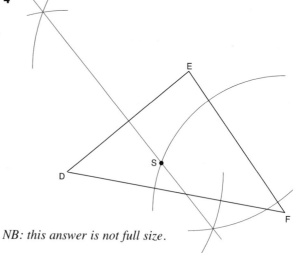

NB: this answer is not full size.

    3

5 $t = 3.2$ **4**

Examiner's tip
It is important that you show all trials and their outcomes.

6 a $10 < t \leqslant 15$ **2**

   b   14.5 hours   **4**

Question	Answer	Mark

Examiner's tip
Remember to multiply the frequency with the midpoint in part (b) and then divide by 60 i.e. the sum of the frequencies.

7 a $\dfrac{3.4}{7.2} \times 8.7$

$= 4.1\,\text{cm}$ **3**

b $\dfrac{13.7 \times 7.2}{10.6}$

$= 9.3\,\text{cm}$ **2**

Examiner's tip
Give answers to a sensible degree of accuracy.

8 $p = \sqrt[3]{16 \times 10^{12} \times 2 \times 10^{3}}$

$= 3.17 \times 10^{5}$ **3**

9 $AB = \dfrac{9}{\tan 55°}$

$= 6.3018...$

$AC = \dfrac{9}{\sin 55°}$

$= 10.98...$

Perimeter $= 26.3\,\text{cm}$ (3 s.f.)   **5**

Examiner's tip
To find AC, Pythagoras Theorem or trigonometry could have been used.

10 £420   **3**

Examiner's tip
Remember to divide by 0.85.

11 a 30 year 7 pupils   **2**

b $\left(\dfrac{85}{175} \times \dfrac{50}{120}\right) + \left(\dfrac{90}{175} \times \dfrac{70}{120}\right) = \dfrac{211}{420}$ **3**

12 a $x = 0$ or $x = 1.5$ **2**

b $x = -1.4$ or $x = 2.9$ **2**

c $x = -0.4$ or $x = 2.4$ **4**

13 a 6.25×3.765
$= 23.53125\,\text{cm}^2$ **3**

b $\dfrac{7.455^2}{6.35}$

$= 8.752287402\,\text{cm}$ **3**

Question	Answer	Mark

14 Area of segment =
sector area − area of triangle

$= \left(\dfrac{40°}{360°} \times \pi \times 6^2\right) - \left(\dfrac{1}{2} \times 6 \times 6 \times \sin 40°\right)$

$= 0.996\,\text{cm}^2$ (3 d.p.)   **6**

Examiner's tip
Answer this question systematically and show all steps in your working.

15 a $\dfrac{1}{2} \times (2x + 1) \times (x + 5) \times \sin 30° = 10$

$(2x + 1)(x + 5) = 40$ $\left(\sin 30° = \dfrac{1}{2}\right)$

$2x^2 + 11x + 5 = 40$

$2x^2 + 11x - 35 = 0$ **4**

b
$x = \dfrac{-b \pm \sqrt{b^2 - 4ac}}{2a}$

$x = \dfrac{-11 \pm \sqrt{11^2 - (4 \times 2 \times -35)}}{4}$

$x = \dfrac{-11 \pm \sqrt{121 + 280}}{4}$

$x = \dfrac{-11 \pm \sqrt{401}}{4}$

$x = \dfrac{-11 + 20.02}{4}$ or $x = \dfrac{-11 - 20.02}{4}$

$x = 2.256$ or $x = -7.755$

since x represents a length:

$x = 2.26\,\text{cm}$ (2 d.p.)   **3**

Examiner's tip
Even if you cannot do part (a), part (b) is simply solving a quadratic equation by using a formula.

16 $42 \times (1.013)^{20}$
$= 54.4$ million   **3**

Examiner's tip
The use of a multiplier is important when working out the answer. Do not attempt to work it out for each year!

17 $x = 3, y = -7$
$x = -1.8, y = 7.4$ **6**

18   Calculate BD

$$BD = \sqrt{(12.6)^2 + (14.2)^2 - (2 \times 12.6 \times 14.2 \times \cos 75°)}$$

$$BD = 16.36 \, cm$$

$$\frac{\sin A\hat{B}D}{17.3} = \frac{\sin 57°}{16.36}$$

$$\sin A\hat{B}D = \frac{\sin 57°}{16.36} \times 17.3$$

angle ABD =   62.5°

                                                6

Examiner's tip
This is an example of a multistep question. First find the
length of BD by using the cosine rule and then the size of
angle ABD by using the sine rule.

---

19   **a** $n^2 + (n + 1)^2$

$n^2 + (n^2 + 2n + 1)$

$= 2n^2 + 2n + 1$

$\underbrace{2n(n + 1)} + 1$

Always even   **3**

$2n(n + 1)$ must always be even as it
is a multiple of 2, then $2n(n + 1) + 1$ will
always be odd.

   b $(2n + 1)^2 - (n + 2)^2$

$(2n + 1 + n + 2)(2n + 1 - n - 2)$

$= (3n + 3)(n - 1)$

$= 3(n + 1)(n - 1)$ **3**

Examiner's tip
Use the difference of the squares!

Total 100 marks

• •

HOW TO ASSESS YOUR GRADE

The grid below suggests grades that you might have expected to achieve with different scores on these papers.
The marks are combined from paper 1 and paper 2 and are out of 200. No account has been made of the
coursework marks. It is an indication only and does not imply that this is the grade you will receive in the
real examination.

Grades D and below are not awarded on Higher Tier.

A*	162–200
A	124–161
B	88–123
C	46–87

Index

Index